PRINCIPLES
OF MECHANICAL DESIGN

McGraw-HILL **T**ECHNICAL
EDUCATION
SERIES

McGRAW-HILL TECHNICAL EDUCATION SERIES

Norman C. Harris, Series Editor

Introduction to Electron Tubes and Semiconductors · Alvarez and Fleckles

Data Processing: A Text and Project Manual · Cashman and Keys

Architectural Drawing and Planning · Goodban and Hayslett

Calculus with Analytic Geometry · Green

Introductory Applied Physics, 2d ed. · Harris and Hemmerling

Experiments in Applied Physics · Harris

Elementary Mathematics: Arithmetic, Algebra, & Geometry · Hemmerling

Pulse and Switching Circuits · Ketchum and Alvarez

Introduction to Microwave Theory and Measurements · Lance

Microwave Experiments · Lance, Considine, and Rose

Electronics Mathematics · Nunz and Shaw
    Volume I: Arithmetic and Algebra
    Volume II: Algebra, Trigonometry, and Calculus

Principles of Mechanical Design · Parr

Electrical Fundamentals for Technicians · Shrader

Specifications Writing for Architects and Engineers · Watson

(Other volumes in preparation.)

PRINCIPLES
OF MECHANICAL DESIGN

ROBERT E. PARR

Assistant Professor of Aerospace Engineering
NORTHROP INSTITUTE OF TECHNOLOGY

McGraw-Hill Book Company

NEW YORK   ST. LOUIS   SAN FRANCISCO
LONDON   SYDNEY   TORONTO
MEXICO   PANAMA

PRINCIPLES OF MECHANICAL DESIGN

Library of Congress Catalog Card Number: 69-13611

07-048512-7

11 12 KPKP   8 3 2

PREFACE

The purpose of this book is to prepare the reader to perform the exacting work of mechanical design. The book is suitable for use in technical institute and junior college curriculums, such as mechanical engineering technology, machine design, and tool design. It is also appropriate for use in curriculums leading to a bachelor of science degree, where a limited time is to be devoted to mechanical design, such as aeronautical engineering and industrial engineering.

In addition to being a text on the design of machine elements, the book is intended: to serve as an aid or guide in the complex process of design; to bridge the gap between courses in drafting, manufacturing processes, mechanics, and strength of materials (it is assumed that the reader has a basic understanding of these subjects) and the actual work of mechanical design; to aid in developing the background upon which decisions are based. In other words, it is intended to facilitate the transition from student to practical designer.

In the first six chapters the design process is analyzed, materials and manufacturing processes are discussed from the designer's point of view, and many other practical considerations that are important factors in actual design are discussed. Teaching design to college students for the past twelve years has convinced the author of the desirability of including this analysis and these discussions, to aid in developing the proper perspective and the ability to produce a practical solution to a design situation. These chapters are based on over ten years of experience in industry as a design engineer and manufacturing engineer.

It may be well to give the reasons for the manner in which the design of machine elements is presented. The most frequently used elements are treated in detail, and examples of problems pertaining to them are given, but in addition, many other items are discussed briefly rather than ignored. This approach is intended to develop an appreciation for the magnitude of the mechanical design field, to make the reader aware of these less common elements, and to create an interest that will result in some additional individual study.

Enough information to work the problems at the end of the chapters and to complete the projects in Chapter 19 has been included. However, no attempt has been made to present extensive design data, because of the space limitation and because such information is readily available in handbooks, manuals, and catalogs.

I wish to acknowledge the competent assistance of my wife, Lucy, in the preparation of the manuscript.

Robert E. Parr

CONTENTS

LIST OF TABLES

PRINCIPLES
OF MECHANICAL DESIGN

1

INTRODUCTION

1-1 WHAT IS MECHANICAL DESIGN?

Engineering is one of the most important activities of man. The pyramids of Egypt and the aqueducts of ancient Rome are early engineering accomplishments. Engineering produced the steam engine and the machines it powered, which allowed the average man to have some of the things that had been available only to the rich. It was engineering that achieved the airplane, fulfilling one of man's oldest dreams. And it is engineering that has made it possible for man to go to the moon. Engineering has the responsibility of providing man with that which makes his life easier and more interesting.

Design is a creative activity, and it is not an exaggeration to say that the creative efforts of man have produced that which most distinguishes him from other forms of life. Creativity as it applies to engineering is the ability to conceive basic innovations, to perceive in a situation those problems that can readily be solved, to devise solutions to new problems, and to combine familiar concepts in unusual ways. Thus, *design is the creative part of engineering*. The word "design" will hereafter be used for this aspect of engineering and "designer" for one who engages in this activity.

There are some activities that may be classed as design which are outside the scope of this book — for example, the architectural and structural features of buildings, civil engineering projects such as highways and dams, the external configuration of ships and aircraft, and electronic equipment. Thus the title *Principles of Mechanical Design*.

1

The end result of the design process will hereafter be called the product. Many features desired in a product are incompatible; thus, it is readily apparent that design involves making decisions and compromises based on careful consideration of many factors. It also becomes apparent that the designer must continuously exercise judgment, and that faulty judgment in only one instance could invalidate the whole design.

Design is a large field and it is becoming larger very rapidly as discoveries are made, as inventions are produced, and as new needs arise. These inventions or discoveries of themselves do not benefit man, for it is only when they serve as the basis of a design or are incorporated into a design that man derives a benefit from them. It is also appropriate to point out that a new design by itself is of no real significance unless it clearly fulfills a need and is produced in an appropriate quantity.

A designer is concerned with such problems as the geometric arrangement of components, the effect the motion of one part has on the motions of those associated with it, and the effects of forces. He is also concerned with the properties of materials, with the capabilities and limitations of manufacturing processes, with human capabilities and limitations, and with economic matters. A designer must also possess a highly developed ability to present his ideas concerning complex engineering problems in such a manner that they are readily and clearly understood by other technical personnel.

Again we shall try to answer the question: What is mechanical design? Mechanical design is the application of many of the principles of science and technology in the creation of a product and the consideration of the various factors that affect its production and use. The product shown in Fig. 1-1 is the result of the skillful application of the principles of mechanical design.

1-2 WHERE DOES DESIGN FIT?

The engineering involved in producing a product may be placed under one of two broad classifications: product engineering or manufacturing engineering. Product engineering is responsible for all the functions that are required to bring the product into being and to describe it completely by the use of drawings and specifications. Included are preliminary design, kinematic design and analysis, analysis of forces and loads, detail design, stress analysis, construction of models and mock-ups if used, construction and tests of prototype, redesign as required, and preparation of final drawings and specifications. Manufacturing engineering is responsible for all the functions that are necessary to produce, in the required quantity, a product that fulfills the requirements set forth by product engineering. This includes planning the manufacturing processes and their sequence; planning the tooling; designing, making, and proving the tooling; ordering parts and

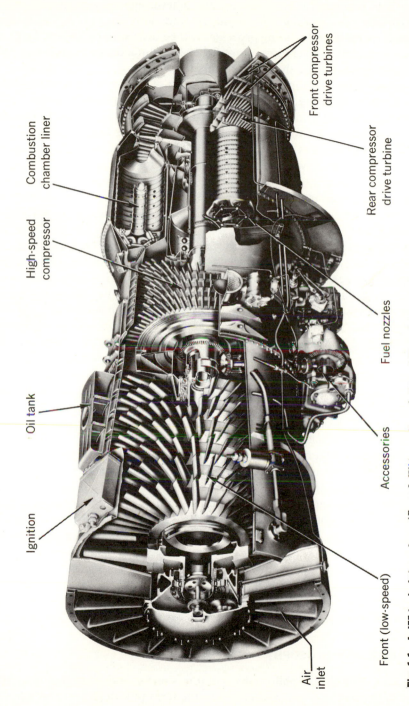

Fig. 1-1 A J75 turbojet engine. *(Pratt & Whitney Aircraft, Div. of United Aircraft Corp.)*

Front compressor drive turbines

Rear compressor drive turbine

Combustion chamber liner

High-speed compressor

Oil tank

Ignition

Fuel nozzles

Accessories

Front (low-speed) compressor

Air inlet

materials; planning and scheduling manufacturing operations; inspection and testing; preparation for shipment; and shipment to the customer. Though many manufacturing processes will be discussed in the light of their effect on design, the very important field of manufacturing engineering will be regarded as outside the scope of this text.

In addition to being concerned with most of the items listed as activities of product engineering, a designer is concerned to a degree with many other problems, including some that at first thought would appear to be strictly business matters and therefore of no concern to a designer.

Research involves considerable thought and the consideration of many factors and often serves as the basis of a product. Thus, it might be concluded that research and design are basically similar; but this is not true. Research is directed toward a generalization, whereas design is directed toward a detailed practical solution to the problems of a particular situation.

Development is an activity that could be confused with research. It differs basically in that it is concerned with the details of things, often with a particular design. Because of this, a designer is at times involved in development. It should not be concluded that a designer is not interested in research, for research provides much of the information essential to his work; it is just that he is not engaged in research.

The designer is at the focus of the conflicting desires of various departments: the sales department wants a product that has appeal and fulfills the need of both customers and prospective customers; the production department wants the product to be easy to manufacture and assemble; the management wants the product to be profitable; and everyone wants the designer's work completed immediately after the decision is made to produce a new product.

There are, of course, various levels of design, some projects requiring more experience and ability than others, but the basic concepts and methods are the same for all levels.

An understanding of the relation of drafting to design is important to an appreciation of the position of design. Drafting is the preparation of drawings that serve as the means by which the engineering department exactly describes the various parts and assemblies that make up the product, and this requires some ability in graphic science. A designer also makes considerable use of graphic science, which is most conveniently done at a drafting table. But to say his work is drafting would be the same as saying the work of an author who uses a typewriter is merely typing.

1-3 DESIGN PHILOSOPHY

One definition of philosophy is that it is one's total attitude toward life. It is in the sense of basic attitudes that the term philosophy is used here —

Fig. 1-2 An industrial hydraulic pump. (*Vickers Inc., Div. of Sperry Rand Corp.*)

attitude toward the engineering profession in general and toward design in particular.

In addition to an individual's own design philosophy, there is a unique design philosophy associated with the particular field in which he is working. Such philosophies of design are rarely recognized and very rarely stated. Nevertheless, this underlying philosophy, though difficult to express, has an effect on the manner in which design work is done and a considerable influence in establishing the criteria upon which judgment is based.

An appreciation of the effect of design philosophy can be gained from a comparison of two high-quality variable-displacement pumps produced by the same manufacturer. They have almost equal displacement per revolution. Figure 1-2 is an industrial pump; Fig. 1-3 is an aircraft pump. The rated output in horsepower of the industrial pump is about one-fifth that of the aircraft pump, and it weighs approximately 3 times as much. Thus, the weight per horsepower of the industrial pump is 15 times as great.

Consistent with the philosophy of a particular field, each of several designers will conceive of the design problems in a situation and the possible solutions differently because of their design philosophy and experience. The designer's philosophy will have a considerable effect on the quality of his work and on how rewarding it will be. If he regards design as an opportunity to use creative talent, to demonstrate ability to deal with complex

Fig. 1-3 An aircraft hydraulic pump. (*Vickers Inc., Div. of Sperry Rand Corp.*)

problems, and to exercise judgment in making decisions involving many considerations, greater success will result than if he thinks of design as concocting something to do a given job in spite of the frustrating difficulties that are encountered.

Thus, it becomes apparent that there is no such thing as "the philosophy of design" with which the beginning designer can become familiar and which will provide guidance. But realizing this will relieve him of much unnecessary concern when he becomes aware of various design philosophies.

An element common to all successful design philosophies is: a designer does what is best for the product, not what is easiest for him.

1-4 CHARACTERISTICS OF A DESIGNER

A successful designer must have the character traits which are desirable for any person in a position of responsibility. In addition, he must have other qualities which are essential only for a designer. No attempt will be made to discuss these in the order of their importance.

Initiative is important, because it is often hard to get started on the real work of design — the tendency is to wait for inspiration. Persistence is necessary, for there is always the opportunity and often an excuse to let things slide or to coast, thinking that a better job can be done some other time, when conditions are better, when there are not so many things that should be done, or after a good night's sleep. The designer must have the ability to distinguish between the essential and the nonessential, to analyze carefully and to attach the proper degree of importance to various factors when evaluating a design. He must be able to concentrate intensely for long periods. He must have the ability to combine familiar elements and ideas in unfamiliar ways to solve new problems.

A designer must be able to visualize parts in three dimensions, and to visualize the forces acting on them; also to visualize assemblies of parts and to an extent the motions of mechanisms. He must be conscientious, able to look at his work critically, and when it is seen to be inferior to another solution that comes to mind, to change his design rather than accept what he has done and then design around it and in so doing introduce an unnecessary compromise. He must have a well-developed and controlled imagination. He should tend to be dissatisfied with things as they are.

It is essential for a designer to have an understanding of fundamentals, not merely a collection of facts and equations. He should be aware of a great many things, even though they do not appear to have anything to do with design, for it is often from such a broad background that ideas come which provide the best solutions to design problems. He must be able to keep in mind several possible solutions simultaneously and to work with information even though he does not know what the outcome will be. He should be willing to take a chance, not in the sense of recklessness, but in the sense of being willing to put time and effort into the solution of a problem even though there is no assurance that it will be a success. He must be a nonconformist to the extent that he does not go along with a solution or an idea just because it is always done that way, but he must not be a nonconformist in the sense that his design will be different for no better reason than just to be different. He should be stubborn in the sense that he is not easily discouraged, but not in the sense that he will continue with a solution he is convinced is inferior just because it was his original idea and he will prove that it will work. The ability to get along well with people is important, because a designer must work with people in other fields in producing a design, and if the product is complex he will have to work with other designers, as he will be given only a portion of it to design.

1-5 DEVELOPING DESIGN ABILITY

In the foregoing section, some of the important characteristics of a designer were mentioned. The question facing the beginning designer is: can he acquire these characteristics and, if so, how? Many of these characteristics are dependent on the basic character of the individual, while others are a form of talent. Some people have them to a greater extent than others; nevertheless, for most of those who would seriously consider becoming a designer, these characteristics can be developed, if sufficient sincere effort is expended. As a beginning, the following practices are recommended:

1.   Develop the ability to visualize objects in three dimensions from a drawing in two dimensions; not just to provide the third view mechanically but to "see" the object mentally from several points of view.

2.   Work at explanations through the use of spoken words and sketches and then by written words and sketches.

3.   Acquire the habit of analyzing mechanical things — watch them work, take them apart — to learn how various problems were solved.

4.   Study engineering drawings, in the same manner as a textbook is studied, to learn design details.

5.   Develop problem-solving ability — not only for problems involving mathematics, but for problems involving mechanical things, such as what can be done when a wrench is too big for the nut that is to be loosened and the only other item at hand is a hacksaw blade.

6.   Work at developing a memory, not for numbers and equations, but for the approximate relation of one factor to another and for mechanical concepts, such as how various motions are accomplished and methods of locking an adjustment.

7.   Develop the habit of being thorough — check to be certain everything that should be present is included, that all that is required to be done has been completed, that nothing has been overlooked, and that those errors that can be discovered by careful checking have been found and corrected.

Attention should be called to the fact that there is not an equation or rule that is used to make each of the decisions involved in producing a design. These decisions require judgment which can be developed only by exercising it, and to exercise judgment a designer must first have the background upon which to base this judgment. It is the purpose of this text to aid in developing this necessary background, in addition to providing an understanding of the principles involved in the design of machine elements.

At this point it may be well to mention a few reasons for becoming a designer. The more obvious reasons are: interesting work, good pay, pleasant environment, and opportunities for advancement. For many, the unique sense of achievement which can be experienced only by a designer who sees that the product he created satisfies the many requirements specified is the most important reason for his continued contributions in the field of mechanical design.

2 THE DESIGN PROCESS

2-1 INTRODUCTION

A designer is not given a set of carefully stated problems, the solution of which will constitute the design he is assigned to produce. He is instead given a situation. Even in the case of a relatively simple situation there are a number of possible solutions, and a beginning designer should realize that they are not equally desirable and that the best is not obvious. Also, he should realize that not all the solutions can be found, and that even if they could, they could not all be developed sufficiently to be evaluated. Nevertheless, a design must be produced, in an appropriate time, that satisfies all the conflicting needs and interests involved in the situation. The amount of time that is appropriate depends on the complexity of the situation and on the financial potential involved.

A beginning designer assumes, or hopes, that there is a method used by experienced designers that he can learn which will facilitate dealing effectively with what appears to be a formidable task. An experienced designer does have a method for creating and developing an acceptable design. But it is not as simple as just learning *the* method, for as shown in the last chapter, there is not a single design philosophy. And because of the influence philosophy has, there is not just one method that is used by all designers. Nevertheless, the method outlined in this chapter will make it possible to discuss the concepts involved in the design process and will serve as a method for the beginning designer until he develops one more suited to his personality, his experience, and the work in which he is engaged.

9

The complexity of the situation will dictate the degree of competence required of the designer, but the design process is essentially the same for either a simple or a complex situation. If the situation is complex, the design process will have to be gone through many times before the detail design of components has been completed.

It must be kept in mind that the separate steps and the sequence referred to as the design process are not intended as an accurate description of what actually takes place, but rather to facilitate an orderly discussion. The activity discussed in several steps is performed simultaneously in actual design. For example, it is necessary to do some of the preparation of Step 2 (prepare to deal with the situation) in order to provide the background that is necessary to the proper execution of Step 1 (become acquainted with the situation).

There will be a tendency to begin the solution a number of times while engaged in the activity of the first four steps; this tendency should be resisted.

2-2   STEP 1: BECOME ACQUAINTED WITH THE DESIGN SITUATION

The importance of this activity is often underestimated. It is essential that everything that may have an effect on the design be given proper consideration, but consideration cannot be given to something of which the designer is unaware.

Just what is it that is needed, who needs it, and why? Where is the product to be used, by whom is it to be used, and how will it be used? When is the solution required, who is to approve the design, and how many units are to be made? Questions such as these should come to mind as the designer acquaints himself with the situation, for the answers will have a great effect on the manner in which the situation should be approached and thus on the solutions that will be produced.

The designer will never be given all the information that he needs about the design situation. Often those who assign a project to a designer do not know what information is needed, or they have the attitude (and it is not inappropriate) that it is part of the designer's job to determine what information he needs and to get it.

There is a great difference in the importance of various design projects; that is, some solutions are worth more than others. Complexity is not a true indication of importance. A good design for a simple product could mean much more to a company which produced a large quantity than would a good design for the more complicated machine used to test the prototype of the product.

The answers pertaining to what the solution is worth should be found as soon as possible to prevent spending too much time, or to prevent the more serious error of not giving the design situation the consideration that the importance of the project warrants. Not all the items that appear in the statement of the design situation are of equal importance. The designer should determine their relative importance so that he can intelligently make the inevitable compromises and direct his effort to the areas where it will do the most good.

Because of the complexity of a project, a designer may be assigned only part of it; therefore, the designer and his design become involved with other designers and their designs. In such a situation, a designer now must consider the effect that the other designs may have on his and the effect that his may have on the others. It would also be helpful to know something of the attitudes, methods, idiosyncrasies of the other designers that pertain to their work.

Part of the process of becoming acquainted with the design situation is to determine the latitude allowed the designer. He must make certain whether his assignment is to provide a means of accomplishing a specified objective, or to design a certain piece of equipment in accordance with a detailed description of what is desired. For example, his assignment could be to provide a means of transmitting motion from one shaft to another which is at an angle to it. In this case, he could choose any of the methods shown in Fig. 2-1. Or his assignment could be to provide for transmitting motion from one shaft to another by means of a pair of bevel gears. It should be remembered that accompanying the greater latitude is a greater responsibility and also that the natural tendency to increase the latitude will often be resisted by others.

A designer should make certain that the design situation is thoroughly understood, including the latitude that he is allowed, so that he will not be guilty of failing to deal with the whole problem.

Also, a part of the process of becoming acquainted is to establish the magnitude, characteristics, and variations of the input that must be provided for. For example, if the product is an automatic screw machine, what type of bar, what maximum and minimum sizes of bar, what maximum length of part, and what kinds of material is the machine to handle? The kind and precision of the output must also be determined. For example, if the product is a rock crusher, the output could be small pieces of rock, the maximum and minimum sizes of which are specified.

If the designer is not familiar with the philosophy of the field in which he is working or of the company by which he is employed, it would be appropriate at this time to put forth some effort to become acquainted with these philosophies.

2-3 STEP 2: PREPARE TO DEAL WITH THE SITUATION

An acceptable design cannot be developed unless the designer is both familiar with the situation and adequately prepared to deal with it. He is

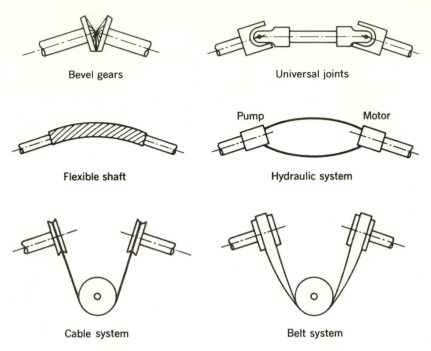

Bevel gears

Universal joints

Flexible shaft

Hydraulic system

Cable system

Belt system

Fig. 2-1 Various ways of transmitting motion from one shaft to another.

not automatically prepared by virtue of a formal education. He very likely will have to review some design principles and learn more about a manufacturing process or a new material.

Knowing how to obtain the appropriate information, and how to use it to properly prepare for producing an acceptable design, is a characteristic of an experienced designer. The problem usually is not that there is insufficient information to serve as a basis for the preparation being undertaken, but that there is too much. The designer must realize that he cannot carefully consider all of it, and that he must select that which appears most promising. He may decide not to make use of some of this information in the design at hand. Nevertheless, he is the only one who can determine whether it may be advantageous to attempt to incorporate it or not, and he must, therefore, acquire the knowledge to enable him to evaluate and decide.

A designer can only produce a design based on his experience and knowledge. And a beginning designer has little of either, especially experience; thus, he must do considerable preparation. It might be concluded that an experienced designer will require little time for this preparation; but this may not be true. For the greater the experience of a designer, the more complex or important are his assignments.

Knowledge of solutions to similar situations is an advantage. If the solution was successful, the same basic method or idea may be used in the new situation; and if the solution was not particularly successful, it will not be attempted again. However, there is a danger here that must be recognized and guarded against. The new situation may not be as similar to the successful situation as it appears. Thus, to use the same idea may result in a poor design, if not a complete failure. Also, the factor that was responsible for the failure in the similar situation may not be present in the new situation, and what may have been a good solution is ignored.

Another reason for the study of solutions to similar situations is that it may prevent overlooking an important factor, or it may bring to mind useful ideas, for ideas beget ideas. The greater the number of relevant ideas and data that are considered, the better the chances of success. However, the designer must remember that the solution is only worth so much and that he is to produce a solution consistent with what it is worth; that is, he must resist the tendency to make a career of the project. Another word of caution is in order relative to analyzing other solutions. It should be kept in mind that the solution may have just evolved, or it may be a modification of something intended for a different purpose, or it may be a rework of an unsuccessful solution, or it may have been the best solution possible at the time and advancements in the state of the art may have taken place since then. The effect of such an advancement is shown in Figs. 2-2 and 2-3. Figure 2-2 is the design of a piston head of a hydraulic cylinder that incorporates a chevron packing. Figure 2-3 is the design of the same piston head using the more recently developed O-ring packing.

It would also be well to keep in mind that relationships, equations, and procedures should not be used without knowledge of their background, that is, the area of applicable application, the assumptions that were made, and the degree of accuracy that is produced.

Information which must be acquired by the designer is found in textbooks, handbooks, periodicals, reports by agencies such as NASA, papers

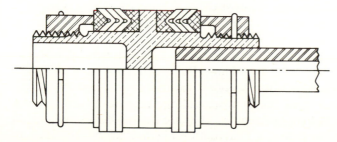

Fig. 2-2 Piston head using chevron packing.

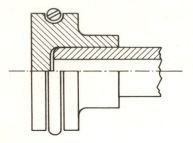

Fig. 2-3   Piston head using O-ring packing.

presented at meetings of societies such as the SAE, and literature supplied by manufacturers and suppliers of materials and equipment. These are the more obvious sources, but there are others, such as other designs and test data of the company by whom the designer is employed, patents, other engineering personnel, even "old hands" in the laboratory or experimental shop. But making use of the ideas and experience of other people must be done skillfully, for there is a tendency for them to become too involved.

2-4   STEP 3: IDENTIFY THE ELEMENTS

The design situation consists of a number of elements which upon careful consideration can be placed in one of several basically different classifications. The designer must recognize the elements in the situation and classify them, for he cannot produce a design by always considering the entire situation. There is no extensive formal statement of the elements in these various classifications. The terms used here for the classifications are, of course, arbitrary and are intended only to facilitate discussion of the process.

1. Relevant problems — These are the elements of the situation that must be most clearly identified and understood, for they are the essential part of the situation and are to receive the greatest attention.

2. Irrelevant problems — These elements are real problems but are not the concern of the designer, because they are not part of his assignment. The designer should be careful to identify such problems correctly so as not to waste his effort.

3. Phantom problems — These are often thought, especially by beginning designers, to be relevant problems. But they are not problems at all. Nevertheless, considerable effort is often expended in "solving" them. They are the result of not fully understanding the situation, or of not clearly defining the relevant problems, or they are created by the designer in his thinking about the situation. The nature of these problems makes it difficult to see them for what they are, but being aware that they exist in

most situations should put the designer on his guard and help him to avoid most of them.

4.   Obstacles — An obstacle is a well-defined element of the situation that has all the characteristics of a difficult relevant problem. However, an obstacle differs from a relevant problem in that the designer will try to avoid having to solve it because it is of such a nature that it appears the design will not suffer because it is avoided rather than solved. If an obstacle cannot be avoided without severely compromising the design, it will have to be considered a relevant problem and will have to be solved.

5.   Difficulties — A difficulty is an element of the situation that is real and of concern to the designer, which may not be clearly recognized or well defined. Difficulties must be overcome, not ignored. Some will be overcome by modifying the relevant problems, some will be found to be irrelevant problems or phantom problems, some may be overcome by redefining an obstacle, and some will be overcome only by making them obstacles or relevant problems.

6.   Hindrances — Associated with any design situation will be things that hardly justify being called an element but they can have a considerable adverse effect on the design. For example, information required to define a problem is known to exist, but it is not readily available. This is a hindrance, and it will have an effect on the design if the problems are modified in a manner such that this information is not required. At times, considerable work is required to overcome a hindrance, and because this work is often uninteresting and annoying there is a natural tendency to avoid it. A designer must resist this tendency to compromise his design.

This discussion of the elements, using such words as problems, obstacles, and difficulties, should not be thought of as discouraging, for successfully dealing with these elements is responsible for much of the interest and satisfaction a designer experiences in his work.

In this search of the situation for the relevant problems, questions about the objectives may come to mind. Answers to these questions must be found by carefully reconsidering the statement of the situation that was made at the time that the assignment was given, or by asking for clarification from the person making the assignment. A designer should not in this instance make an assumption.

It is essential that this searching of the situation be done carefully, for if it is not, there is a probability that a relevant problem will not be recognized. This may result in an unsatisfactory design and will at least cause trouble at a later stage in the design process.

In analyzing the entire situation in an effort to identify the elements of the situation and eventually the relevant problems, it is necessary to have some idea of what the design will be, though it will be a rough idea or general concept. It must be remembered that at this point the interest is not in solutions but in isolating the relevant problems that could have to be solved, and that the general concept of the design is merely to aid in finding these problems. The designer should also be aware that there will be a powerful, though unconscious, tendency to see in the situation that which he wants to see, that which is familiar, that which he thinks will be easy to solve, or that which will be interesting to develop. He must also guard against stating as a problem something impossible of solution and then conceiving other problems on the basis that the first problem will be solved.

2-5 STEP 4: ANALYZE THE ELEMENTS

The fourth step is to analyze the elements that have been identified in the last step, especially the relevant problems. There is a tendency to take the attitude, "I will worry about the problems when I come to them." This is a very poor attitude for a designer to take at this time, for when an important factor is discovered after the design is partly completed, it may be provided for by making compromises that would not have been regarded as acceptable when the design was in the earlier stage. Making appropriate compromises is part of the design process, but a designer should not begin with the thought that compromises can be made to compensate for failing to make an effort to find all the factors that may have an important bearing on the design. There is the ever-present danger than an important factor will be overlooked. Thus, a designer must do all he can to prevent this. And a sincere attempt to think of everything that may affect his design is the best thing he can do.

There are, undoubtedly, ideas and information that will facilitate producing a design that is superior to that which is first conceived. But if these ideas and this information are not introduced very early, it is likely that they will not be given consideration at all. Thus, they not only will not be used to advantage, but their omission may adversely affect the design.

The tendency to engage in solving the problems must be resisted at this time, though any ideas should be noted so that they will not be lost. But time and effort should not be spent in trying to develop them. Care should also be taken that a solution is not suggested in stating and defining a problem.

A great deal of interdependence exists among the elements of a situation, and considerable care and patience are required in order to define the relevant problems clearly and in order not to eliminate a possible solution by the process of analyzing and defining.

At this point it would be well to call attention to an important but simple truth: many problems are difficult merely because they have not been properly and fully defined.

Upon careful examination, relevant problems are found to consist of requirements, limitations, and restraints. Requirements are of the nature of what in detail is to be accomplished, that is, the conditions or arrangements that are to be provided. Limitations are largely imposed by the laws of physics. Restraints are of a more arbitrary nature and come from various sources, such as from individuals or groups within the company or from customers.

A designer is interested in limitations and restraints, for they form the boundaries of the area in which he must work to satisfy the requirements.

Requirements and restraints should not be accepted by the designer without question, for some, on analysis, will be found to be more wishes or hopes than essentials, and if retained they may influence the design adversely. Also, careful analysis and study may prove that what appears to be a definite limitation is just a characteristic of the solution to a similar problem or is due to the manner in which the design situation was presented.

Careful analysis of the relevant problems will also indicate which features of the design are less flexible in the sense of the number of possible solutions that would prove acceptable, as well as which features are more critical in the sense of the greater quality or precision that will be necessary in providing for them. It is important that the inflexible and critical features be recognized and that the extent of compromise they can tolerate be determined.

The relevant problems should finally be analyzed to make certain that those which appear difficult cannot be avoided. For those that must be dealt with, the degree of confidence that an appropriate solution can be produced should be determined. If there is not a sufficiently high degree of confidence to justify beginning the design, the entire situation should be reconsidered.

2-6 STEP 5: CREATE THE DESIGNS

This step in the design process is the most important. It is the step for which all the previous steps were preparing. All future steps are concerned merely with developing that which is done in this step. This is when the solution to the relevant problems is undertaken, when the design is created.

It is essential that the designer initially expend considerable effort in dealing with the entire situation; that is, he must concern himself with the situation in its most basic form. As was mentioned earlier, it will be necessary to go through the design process a number of times, the first time dealing with the problems associated with the original situation, the other times dealing with the problems that are associated with the situations created by the solutions to the original problems.

An effort must be made to deal with the entire situation. Otherwise,

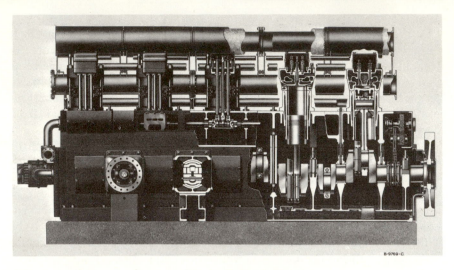

B-9769-C

Fig. 2-4   An engine-compressor unit. (*Ingersoll-Rand Co.*)

though an adequate solution to all the problems is developed, the design as a whole may not be as effective as it could have been in providing for the real problems inherent in the situation.

An excellent example of dealing with the entire situation is shown in Fig. 2-4. A compressor was required for use in a *natural gas* transmission pipeline. The single unit shown combines a 4-cylinder compressor and a 2500-hp 10-cylinder engine that uses *natural gas* for fuel.

In the previous steps, the activity was of the nature of analysis, of breaking down a situation into its elements, of looking for problems. Now, the activity is of the nature of synthesis, of combining elements, of assembling individual solutions into one grand solution. This involves mentally arranging and manipulating not only images of the components but images of forces and motions. The success of this activity is to an extent due to the aptitude of the designer and the methods he uses, but it is to a much greater degree dependent on his knowledge and the effort he puts forth.

Producing a really new design, that is, exercising creativity, requires the destruction of old ideas and well-established ways of doing things. It is difficult for some people to destroy things so dear, and thus they do not become proficient designers. When attempting to produce a superior design, a designer is faced with a very real conflict between (1) the desirability of using proved concepts or elements and running the risk of producing a design that is not making the most of recent developments in the state of the art and (2) the desirability of using ideas that are new or unusual and running the risk of producing a design that suffers from some factor being misjudged or overlooked.

With this emphasis on creativity and the unique, a word of caution is appropriate. If a situation does not require radical departures from accepted concepts and methods, they should not be used merely to produce a design that is different. Nothing is gained by doing so; additional risk is involved, and it is therefore not a good design practice.

It cannot be emphasized too strongly that the tendency to evaluate ideas and solutions must be resisted at this time. There is also the tendency for a designer to be pleased with his ideas, and he may become committed in his thinking to his first good idea. This will effectively block the flow of ideas, one of which may be superior to the first.

Whenever a situation is such that decision A is dependent on decision B and decision B is, in turn, dependent on decision A, as is almost always the case in design, a start must be made. This beginning will not produce a solution, but it may reduce the number of variables or reduce the magnitude of some of the variations. In other words, this faltering and tentative beginning will provide the designer with some rough approximations that can serve as the basis for another attempt. This process of modifying and refining approximations is a very important method when working with complex situations.

In the initial stage of producing a design, especially when dealing with the entire situation, it is best not to give any thought to practical considerations; that is, a designer should not divide his attention when doing the preliminary work in creating a design. It is also important to concentrate only on those features that are essential, critical, or inflexible, ignoring those that are merely desirable.

When attempting to create a design, it is often helpful to describe the function or purpose of a feature or component of the design in two or three words, then to try to think of various alternatives based only on the two- or three-word description. Another method is to think of ways — even fantastic ways — of achieving the objectives by letting the imagination run wild. This may result in an idea that is not fantastic or in a way of modifying another idea, and will at least help to get the designer's thinking out of a rut. Another method is for the designer to begin asking himself an endless series of questions about the various features of a tentative design, such as: Why is this necessary? What is the purpose of this? Can something else be used in place of this? Why do it this way? How can this be done some other way?

In using any of these methods, a record should be made in the form of a written note or sketch of each idea generated. There are two reasons for this: to prevent the idea from being lost and to eliminate concern about keeping in mind the ideas that have been produced. Relieved of this concern, the mind is better able to work at producing another idea.

It is appropriate to create several different designs and develop them only to the point where the basic concept and arrangement can be clearly understood. This is so that they can be evaluated and compared. To do this, it may be necessary to replace the sketches with drawings in which greater care is given to the configurations, sizes, and positions of the various components.

2-7 STEP 6: EVALUATE THE DESIGNS

If the work involved in Step 5 has been well done, there exist a number of possible designs. And now it is necessary to evaluate them in order to select the one that is to be completed. If one of the designs is superior to the others, or if they are all equally desirable, there is no need for evaluation. This is seldom the case, and even if it appeared to be so, in a particular instance, subjecting them to a process of evaluation would still be a good idea, for what appear to be their relative merits may not be substantiated by a careful comparison.

It is no more possible to make a perfect evaluation of the designs than it is to produce for consideration all possible solutions to a complex situation. Also, there are many uncertainties involved in making such an evaluation. Nevertheless, an evaluation must be made, and it must be made with care, for no amount of skill in developing a design can compensate for a poor choice.

The following will serve as a guide in making the evaluation:

1.   The designs should be compared only on the basis of relevant items, such as cost, resistance to vibration, or reliability.

2.   Only those features of the design should be compared that satisfy the requirements in varying degrees, or that are different, such as gears in one design and belts in another.

3.   In establishing the relative importance of the items that serve as the basis for comparison and the relative quality of various design features, sufficient effort must be expended to assure the validity of the values assigned.

4.   Effort should not be wasted in evaluating a design that is obviously inferior to the others that are being considered.

5.   Care must be exercised to prevent choosing a design that is superior theoretically but is inferior practically.

6.   The designs must be compared on the basis of how well they satisfy the requirements that were part of the original statement of the design situation.

7.   Consideration should be given to the implied requirements. These may have as their source the philosophy of the field in which the designer is working or the company by whom he is employed.

8. Consideration should be given to the criteria, as nearly as they can be determined, by which the final design will be judged.

9. Thought should be given to the details of the designs to determine whether the problems associated with the details of one design may be as easily dealt with as those associated with another design.

10. Any unusual material or manufacturing process required by a design should be considered.

11. It should be remembered that one design may be more sensitive than another; that is, it may require more precise parts and adjustments, or it may be more greatly affected by factors in the environment such as heat or dust. This sensitivity may require greater maintenance or a reduction in reliability.

The purpose of Chap. 5 is largely to provide the background upon which to base the judgment required in instances such as evaluating alternatives. The evaluation process should be carried out in a systematic manner to reduce the possibility of overlooking an important item. A procedure similar to that discussed in Sec. 6-7 would be appropriate.

2-8 STEP 7: REWORK THE SELECTED DESIGN

Now that a design has been decided upon, the tendency is to begin the detailed development of it immediately, for there is always the need to complete a project as quickly as possible. This tendency should be resisted. Instead, an effort should be made to modify the selected design. The designs that were rejected should be carefully analyzed with the thought of incorporating any of their desirable features.

The statement of the design situation and the pertinent design considerations should be reviewed in conjunction with analyzing the rejected designs. When a design is almost completed, the designer may see changes that would be desirable, but there may not be time to make them, or they may not be important enough to justify starting over. The activity of this step is largely an attempt to see these changes before the detailed development is begun. As a design develops, it becomes less abstract and more concrete, less general and more detailed, and thus less readily modified.

When attempting to rework the selected design, a feature that is not essential but is desirable may be discovered, and it may appear that this feature can be incorporated without unduly compromising other features. This is the purpose of this step. But the tendency is to let the incorporation of this desirable feature become an inflexible requirement that may greatly complicate the development of the design and even necessitate compromise that its incorporation does not justify. Being aware of this tendency, it can be guarded against, and the design will be improved by those features that are incorporated.

It is appropriate for the designer to ask himself, with respect to the selected design, a number of questions, such as: Would it be better to make the design more flexible or less flexible? Should the life be increased or decreased? Would it be well to improve the appearance? It is true that when this design was selected consideration was given to such items as life and flexibility, but this is the time to consider those changes that cannot be made later without starting over from the beginning with the detailed development.

It would also be well to give consideration to materials and manufacturing processes. This should be done so that the detailed design can be produced with the capabilities and limitations of the materials and processes in mind. In other words, now is the time to make decisions such as: Is the part to be made from an iron casting or an aluminum forging, or should it be an aluminum die casting? Such decisions greatly affect the detailed design, and it is much better to make them early, when the design can be modified to accommodate them, than later, when accommodation may require less desirable compromises.

Sometime before the detailed design is begun, considerable thought should be given to the fact that in the theoretical kinematic arrangement all surfaces, dimensions, and alignments are perfect; also that in a practical mechanism wear, tolerances, and deflections due to loads are unavoidable. This difference between the theoretical and the practical may be the cause of

Fig. 2-5 An IBM Selectric ® typewriter. (*International Business Machines Corp.*)

Fig. 2-6   The typehead used in the IBM Selectric.
(*International Business Machines Corp.*)

unnecessary stresses, excessive wear, unrealistic tolerances, or the failure of
the mechanism to produce motions in accordance with the theoretical
scheme. In providing for these differences between the theoretical and the
practical, the following should be considered: adjustments to allow for
tolerances and wear; self-aligning features, such as self-aligning bearings, to
allow for deflections and misalignment; and the elimination of any unnecessary means of locating or aligning the various members in order to
prevent inducing unnecessary stresses.

2-9   STEP 8: DEVELOP THE SELECTED DESIGN

The basic concept and general arrangement of the design has been
decided. This is the design that will be developed, and it is in the development of a design that a designer spends by far the greatest part of his time.

A design that required extensive development is shown in Fig. 2-5. Un-
like other typewriters, this one uses the typehead shown in Fig. 2-6. The
carrier upon which the typehead is mounted moves across the stationary
paper. In a conventional machine, the print element moves over 5 in. In
this one, the typehead moves about ¼ in. A conventional machine has a
carriage that weighs over 8 lb. The carrier in this machine weighs about 1 lb.
The lower weight and shorter movement result in a machine that is inherently both faster and quieter. An additional advantage is that the typehead can be quickly replaced with one having a different typeface.

The activities engaged in thus far in the design process have been primarily concerned with the design situation as a whole, that is, with overall design. Unless the design situation is very simple, this process has been rather formal, with notes, sketches, and evaluation graphs. All of this is necessary to deal with a complex situation adequately. But only the basic design has been decided upon, and it is now the source of a number of design situations that may, for convenience, be referred to as second-generation situations. Each of these second-generation situations is to be dealt with in a manner similar to that of the original design situation. The treatment of each will not be as extensive as in the original situation, for they are not as involved. But the entire process will be repeated. This will result in a number of third-generation situations, each of which is subjected to the design process. This continues until the design situation is quite simple, for example, to select a nut. But there are still alternatives, such as a plain nut, a nut with a lock washer, a self-locking nut, or a nut with a cotter pin. And consideration must be given and evaluations made, in order to make the proper choice. Even at this level the design process is being used, though since it is not being used in the formal manner, it may not be recognized.

Many of the third-generation situations could not be identified at the time the original situation was being worked with, for they did not exist. As a designer gains experience, these situations, though not foreseen in detail, cause few major difficulties, because he is always thinking ahead.

As a design develops, there is a decrease in the magnitude of the changes that are made as compromises and to accommodate various features, but there is no decrease in the importance of these changes to the success of the design. At this point it is very important that no changes in the basic design be made unless they are essential. Such changes will require new compromises and may unintentionally introduce factors that will result in a poor design, if not in failure, largely because they were not considered when the evaluation of the design as a whole was made.

There is a tendency for a designer to produce a *fine* mechanism, that is, to include features that are not necessary, or to make some features more elaborate than is warranted. This may be due to enthusiasm, or to the natural desire of a designer to demonstrate his skill, or to some obscure reason. It must be resisted. A designer should keep in mind that unnecessary frills do not improve a design, and that a design that is successful and appropriate is in effect a monument to his skill and judgment.

Practically this entire text is concerned with the activity of this one step. Nevertheless, it would be well to list here some basic concepts:

1. The number of instances where bending or combined stress is involved should be kept to a minimum.

2. The center of resistance of a part should be made to coincide with the line of application of the load that it is to carry.

3.   An unintentional mechanical advantage or inertia force should be reduced to a minimum.

4.   The design should be worked from the inside outward; for example, in a gear case the progression is from the shaft to the housing rather than from the housing to the shaft, which could result in a shaft that would be too small.

5.   The more critical and less flexible features should be worked out first.

6.   The size of the part has an effect on the configuration of the part; that is, a configuration that is best for a particular part is not necessarily best for a part of the same nature that is many times as large.

Generally, when a designer has completed a design, he is dissatisfied with it.   He is sure that he could do a better job if he were to do it over.   This is normal, and not necessarily an indication of a poor design.   If time is available — it usually is not — the designer should be convinced that the improvements are worth the time involved.   It is usually thought that the redesign can be done quickly.   But to do it quickly greatly increases the risk of overlooking something.   This risk should not be taken for only a slight improvement.

2-10   STEP 9: PREPARE TO COMMUNICATE THE DESIGN

When the design has been developed, it may be concluded that the designer's work is finished.   This is not true.   The design exists in the form of the drawing that was prepared to facilitate the development of the design and in the form of some papers upon which the required calculations have been made.   All this is clear to the designer, and likely only to the designer. A design must be communicated to others if it is to fulfill its purpose.   The designer is the only reasonable one to undertake this job of preparing for communication, for he understands better than anyone both the design and all that was involved in its creation and development.

There are two reasons for preparing a design for communication.   One is so that the detailed drawings to be used in manufacture can be prepared — the designer could do this, but this is not the usual procedure.   The other is to obtain approval of the design.   The first reason will be discussed further in Chap. 6; the second will be discussed in this step.

The designer must keep in mind that those with whom he is trying to communicate are much less familiar with the design situation, with the problems involved, and especially with his design, than he is.   What may appear obvious to him is not only not obvious, but may be difficult to comprehend even when pointed out.

Methods of presenting information that are familiar to those to be communicated with should be used, for if there is the added difficulty of an unfamiliar method, the reaction may be negative to avoid the extra effort required to understand, or because of the irritation of the unfamiliar method.

It is human nature to resist change, or the acceptance of something new, or the ideas of another person, and a design usually involves all three. One of the basic causes of this resistance is fear of giving approval to something that may not be successful. The preparation should, therefore, be directed toward providing the following:

1. An understanding of the physical configuration
2. An understanding of the principles upon which the design is based
3. An appreciation of the problems involved
4. A demonstration that all the requirements are fully satisfied
5. An assurance that no limitation has been ignored
6. Evidence that all restraints are complied with
7. Proof that attention has been given to all pertinent design considerations
8. Evidence for the validity of the theoretical concepts involved and for the accuracy of the associated calculations

The preparation should be so well done that all this is easily gained. This is not such an imposing task as it appears. The layout is made more complete by adding sections, views, and details; care is taken when making the original calculations and notes, and they are now arranged in an orderly manner. A model or mock-up may be appropriate; these are discussed in Chap. 6.

If the design must be presented to people who have little design background, it would be well to put greater emphasis on neatness, to make pictorial drawings, and to provide more realistic models. In such a situation, it should be kept in mind that a feeling of confidence in the ability of the designer is largely substituted for a thorough understanding of the design, and that this feeling of confidence is based on things that are not especially important to the person with considerable design experience.

If a restraint that has interfered greatly with producing an appropriate design seems unreasonable, consideration should be given to producing a design without the restraint in addition to the one with it. Both designs are then submitted for consideration. The design with the restraint must be produced as though it were valid, for it may be the case. And, anyway, the designer would not want to be guilty of the charge that he had not really tried to produce a design with the restraint.

2-11 STEP 10: GAIN ACCEPTANCE OF THE DESIGN

This is the final step in the design process and an important one, for if the design is not accepted, the work that has been done is largely wasted;

and a designer does not establish a desirable reputation on the basis of rejected designs and wasted effort. Because of the importance of having his
design accepted, a designer must do all he can to gain this acceptance. For a
design to be accepted, it must be adequately communicated, as discussed in
the last step. But there is more to gaining acceptance than merely com-
municating. The designer must, in effect, "sell" his design. It will be convenient hereafter to refer to the activity involved in gaining acceptance of the
design as "selling" and to the person or persons who must accept or approve
the design as the "buyer."

Selling is difficult for a designer, because his abilities, interests, and
aptitudes are greatly different than those of a salesman. Many good designs
have not been accepted because of a poor job of selling. There is also the
possibility that the buyer may lose confidence in a designer merely because
of his poor performance in selling his design.

A designer is convinced that he has a good design, and it is natural for
him to feel that it should be accepted without his having to sell it. There is
also a natural tendency for a designer to feel that he is finished, and a
psychological letdown takes place. Both of these factors interfere with the
selling of a design. He should be careful not to imply that he is selling the
design on the basis of, "I have done a good job," but instead, should appeal
to the personal interests of the buyer.

The greatest part of a designer's effort in selling should be expended in
selling the buyer. This should be done by carefully explaining the design
and in doing so demonstrating that it was done in accordance with all the
requirements, limitations, and restraints. The explanation should have
skillfully worked into it reasons why suggestions and ideas made by others
could not be incorporated if they were not. Credit should be given to those
whose suggestions were helpful, even if in a general way, and this must be
sincere. Answers to important questions should be given before they are
asked, but the tendency to present all the material assembled to sell the
design must be resisted. Much of it is to be used only if required to substan-
tiate statements that have been made, to provide answers to specific questions
asked, and to demonstrate the validity of various decisions. In making the
explanation, it must be constantly kept in mind that people are generally
reluctant to show that they do not understand, but a presentation so elementary or detailed that it could be an insult to the intelligence of the buyer
must be avoided.

Throughout the selling process the designer must be careful not to say
anything that could be taken as a personal criticism, even of those not
present. He should realize that mannerisms he may have that normally
would be ignored may at this time be annoying and, therefore, an effort
should be made to avoid them. It would be much better for him to help
analyze the design than to defend it. If it is good, the analysis will help
to sell it, but if it is poor, he can only harm his reputation by emotionally

defending it. The impression that he is willing to make changes if it will considerably improve the design is much more appropriate than for him to display the attitude, "Nothing is perfect and this is good enough."

An effort should be made to arrange for selling the design when there is sufficient time to consider it fully and to arrive at a decision. There must be no attempt to use the fact that there is limited time, either for consideration or for revision, to sell a design.

It would be well to call attention to several possible conflicts that may have an effect on the acceptance of a design. First, there may be a considerable difference in the design philosophy and background of the designer and the buyer. Second, there may be a difference between what the designer wants to accomplish on a personal basis (for example, to demonstrate his competence) and what the buyer wants on a personal basis (such as a design consistent with his previous decisions). Third, there may be a personality conflict between the designer and the buyer.

It should not be concluded from this discussion that this process of gaining acceptance of a design is an especially difficult or unpleasant task for the designer. If he goes about selling his design in an appropriate manner, he can have the pleasant experience of receiving an expression of sincere approval from those in a supervisory capacity.

The best time to begin gaining acceptance of a design is when engaged in the activity of Step 1. At this time a designer should ask those who are going to evaluate his final design for suggestions on the general concept or factors they regard as particularly important. This is primarily to create in them an identification with, and a personal interest in, the design that is to be developed.

The above paragraph emphasizes the statement in the introduction to this chapter that the separate steps and the sequence referred to as the design process are not intended as an accurate description of what actually takes place, but rather to facilitate an orderly discussion.

QUESTIONS

2-1 Why is it important to know what a solution is worth early in the design process?

2-2 Why is a designer concerned with the latitude that is allowed in producing a solution?

2-3 Why is a knowledge of solutions to situations similar to the one a designer is working with an advantage?

2-4 List five important sources of information that a designer may use.

2-5 What are the more important elements in a design situation?

2-6 What are several ways in which the element referred to as a difficulty can be overcome?

2-7 Why should a designer not take the attitude, "I will worry about the problems when I come to them"?

2-8 What is the difference between a requirement and a restraint?

2-9 Should a designer give thought to practical considerations in the initial stage of producing a design?

2-10 Why should a written record be made of each idea generated in the process of creating the design?

2-11 Briefly discuss three important basic principles that should be considered in developing the procedure for the evaluation of designs.

2-12 What are some of the ways in which a kinematic scheme differs from the practical mechanism based on it?

2-13 Why should a designer resist the tendency to immediately begin the development of the selected design?

2-14 List four basic concepts that concern the development of a design.

2-15 What is involved in preparing a design for communication?

2-16 Why should special attention be given to the methods of presenting information when preparing a design for communication?

2-17 What are some of the important points that a designer should keep in mind when presenting his design for acceptance?

3 MATERIALS IN DESIGN

3-1 INTRODUCTION

The selection of an appropriate material is part of the design process and is therefore the responsibility of the designer. It is not an exaggeration to say that the success of a design depends on the choice of material, or that the most important design problem is often the selection of a suitable material. The large number of materials available makes the problem of material selection more difficult, but it also makes possible the successful solution of many design problems that could not otherwise be seriously considered.

The importance of selecting an appropriate material is well illustrated by Fig. 3-1. Figure 3-1a shows a cast-iron pump impeller after only 4 days of service in a foundry, pumping a liquid that contains an abrasive. Figure 3-1b is the same part made from urethane rubber, still in the like-new condition after 9 months of service in the same application.

Material is selected to provide those characteristics that are essential to the design and as many as possible of those that, though not essential, are desirable. In making the selection it is important to differentiate clearly between the essential characteristics and those that are merely desirable. Many of the items discussed in Chap. 5 should be considered when selecting a material.

In deciding which material to use, it is obvious that careful thought must be given to the factors that will have a bearing on the successful performance of the part after it is made. Though less obvious, it is not less important to give consideration to the effect that

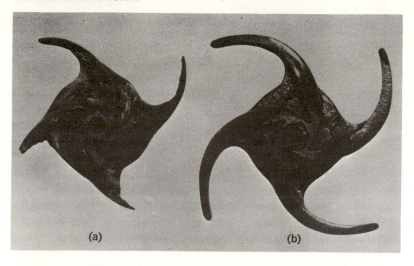

(a)   (b)

Fig. 3-1 The selection of an appropriate material is essential.
(*E. I. du Pont de Nemours & Co.*)

the material will have on the manufacturing processes required to make the part. It often happens that the choice of material dictates the manufacturing processes which must be used. Considerable thought should be given to the question: Will the material being considered permit the required manufacturing processes to be performed at a reasonable cost?

Variations in the composition of materials, such as the small amounts of alloying elements of steel, cause important changes in the characteristics of the material. But it is well to keep in mind that the effect on the material of manufacturing processes and the type of load can be much greater than the effect of variations in composition.

The characteristics that are required of a material vary greatly from one part to another, and the best choice is not necessarily the highest-quality material, for cost is always a factor. Metals are the most common material used in mechanical design because they, more often than any other, will more fully satisfy the requirements. But plastics are being used in ever-increasing amounts, because of their low cost and the special properties they possess.

The proper procedure in regard to material is to determine the characteristics that are most desirable, select the available material which most completely provides the desired characteristics, and then design around this material. Often the choice of material must be based not on the requirements of a single part, but on the requirements of two or more interacting parts. In any event, the selection of a material invariably becomes a compromise, and to make the best compromise requires the exercise of judgment. The purpose of this chapter is to provide the basis for this judgment.

It is not intended to take the place of materials handbooks, though detailed information on some representative materials will be given to permit the selection of a material for a particular application that will be close enough in characteristics and properties so that the final selection will not require revision of the basic design, and to provide the information required by the problems and projects.

As a beginning, the more important characteristics of metals will be briefly discussed from the point of view of the designer and for the sake of continuity. Though the discussion applies specifically to metals, the principles apply to most materials.

3-2 STRENGTH

The strength of a metal is a measure of its capacity to resist forces imposed upon it under given conditions. The force may be steady or rapidly fluctuating, with or without impact, and applied at high or low temperature. The strength of the metal is, of course, affected by these conditions. In one application, a part has acceptable strength if it does not fracture during the life of the part regardless of the magnitude of deformation, whereas in another application a part has adequate strength only if there is a very small amount of deformation under the maximum load. The ultimate strength is, therefore, seldom used in design without some modification. The ratio of the ultimate strength to some other measure of the strength of the metal, such as the yield strength, is sometimes used as an indication of the characteristic of a metal that is desired for a particular application. For example, if it is important that a part adapt itself to a slightly different configuration when subjected to a high stress, or that it provide evidence of excessive load by its deformation but not fracture, a material with an ultimate strength considerably higher than the yield strength would be the best choice.

In any evaluation involving yield strength, it should be noted that, to have any significance, the offset that determines the yield strength must be given. It should also be noted that many of the metals that possess the capacity to deform to a large degree before fracture, as indicated by the ratio of ultimate strength to yield strength, have a very low proportional limit.

3-3 MODULUS OF ELASTICITY

Where the requirement is for a minimum of deflection under load, that is, for a maximum of stiffness, the modulus of elasticity is used to evaluate metals; the higher the modulus of elasticity the stiffer the metal. It would be well to mention that increasing the strength of a basic metal by heat-treating or by adding alloying elements will not increase its stiffness. For ductile metals, the modulus of elasticity in compression is equal, or very close, to the modulus of elasticity in tension, and the modulus of elasticity in shear, also known as the modulus of rigidity, is about 40 percent of the modulus of elasticity in tension.

3-4 DUCTILITY

Metals possessing considerable strength and capable of large permanent deformation before fracture are said to be ductile. The percentage of elongation is most often used as a measure of ductility. A ductile metal has the desirable capacity of reducing the effect of stress concentration by slight yielding, which redistributes the stress. The elongation required to accomplish this redistribution of stress will vary with the type of stress concentration, but it is seldom more than 2 percent. When evaluating materials by use of percentage of clongation, it is important that the gage lengths be equal.

Malleability is closely related to ductility, the difference being that ductility is considered a tensile quality, whereas malleability is thought of as a compressive quality. Malleability is an essential characteristic of metal that is to be rolled into sheets.

3-5 BRITTLENESS AND HARDNESS

Brittleness is that property of a material that is the opposite of ductility; that is, a brittle material will fracture or shatter with very little deformation. This does not mean that a brittle metal has little strength, but that it has a very low percentage of elongation. Brittleness is a characteristic that is seldom required of a metal in design, but it must not be ignored, for it sometimes may accompany a property that is desired, such as hardness.

Hardness of a metal is its resistance to indentation or scratching. For many parts, hardness is very desirable. In general, the harder a metal the more brittle it is and the lower its resilience. For most applications, hardness is required only on the surface; thus, many parts are hardened only on the surface. This is discussed under heat-treating.

The strength of a metal is closely related to its hardness. Therefore, the strength that is desired is sometimes indicated by specifying hardness. Even when the strength is specified in pounds per square inch, it is most often checked by measuring the hardness. The two most common methods of specifying hardness are Rockwell C and Brinell hardness number (Bhn). The approximate equivalents of these two methods are given in Table 3-1.

TABLE 3-1 Approximate Equivalent Hardness Numbers

ROCKWELL C	BRINELL HARDNESS NO.
20	226
25	253
30	286
35	327
40	371
45	421
50	475

3-6 RESILIENCE AND TOUGHNESS

The ability of a metal to store energy without permanent deformation is called resilience. A part subjected to shock loads should be made from a metal that has a large amount of resilience. The magnitude of the resilience of a metal is indicated by the area under the stress-strain curve up to the proportional limit. Thus, a metal that has a high elastic limit and a large amount of strain at its elastic limit possesses a greater degree of resilience than if either the strain or the proportional limit were smaller.

The ability of a metal to absorb energy before fracture is called toughness. A part that may be subjected to an impact load greatly in excess of the load it normally carries and must not break, though the part must be replaced, should be made from a metal possessing great toughness. The toughness of a metal is indicated by the total area under the stress-strain curve. Toughness is affected by the presence of a discontinuity (that causes a stress concentration) to a greater degree in some metals than in others; this is referred to as notch sensitivity.

3-7 EFFECT OF TEMPERATURE

The effect of low temperature on metals is generally to increase the strength, hardness, and modulus of elasticity. There is also a tendency for the metal to become more brittle. This tendency is especially pronounced in the case of plain carbon and low-alloy steels. Thus, if the part is to be subjected to shock loads at low temperatures, these steels should not be used. Steels containing large amounts of nickel have satisfactory shock resistance at temperatures lower than $-300°$F.

The effect of high temperatures on metals is generally to cause a considerable decrease in the strength and stiffness. If a part is subjected to a constant load, there will be a tendency for deformation of the part to increase with time, which is referred to as creep. At room temperature, even if the time is measured in years, this is usually of no practical significance. But if the temperature is increased to several hundred degrees Fahrenheit, this deformation becomes an important factor.

If a part is used in a manner that causes a fixed deformation (such as a bolt used to clamp two parts together) and is subjected to high temperature, there will be a tendency for the initial load to decrease with time. This is called relaxation. Because of the decrease in strength with an increase in temperature, and an increase in both creep and relaxation not only with an increase in temperature but with time, selecting a metal and its appropriate strength for a given temperature and part life becomes a complex problem that requires expert judgment and a thorough study of the conditions. Oxidation and thermal expansion are also factors that may require attention at high temperatures.

3-8 BEARING CHARACTERISTICS

Whenever one mechanical element rests on another and there is a sliding motion between them, a bearing is involved. In this situation, there are a number of factors that must be considered, such as lubrication, temperature, corrosion, and wear, which are discussed in detail in Chap. 12. Though almost any metal can be used somewhat successfully as a bearing, there are metals that are much more suitable for this purpose because: they readily conform to the mating part; they tend to smear across the surface and thus prevent galling and seizing; they readily permit hard particles or grit to be embedded, which prevents scoring and wear. These metals are said to have superior bearing characteristics.

3-9 DAMPING CAPACITY

A rapidly applied load subjects a part to a repeated stress cycle, which if it is not damped can result in a resonant vibration that may cause poor performance or early failure of the part. That characteristic of a metal which enables it to dissipate the energy associated with vibration is known as damping capacity.

In most instances, a high damping capacity is desirable. It is especially desirable in the case of machine frames.

3-10 FATIGUE AND ENDURANCE

A part subjected to many applications of the load may fail even though the stress induced by the load is considerably less than the strength of the metal under static conditions. This kind of failure is known as fatigue failure. Fatigue failure invariably begins at an irregularity in the metal or at a discontinuity due to design which serves to increase the stress at that point. The fatigue strength of a metal is the stress that it can withstand for a specified number of stress cycles.

It seems reasonable that at a stress sufficiently low the number of stress cycles could be infinite. For ferrous metals and alloys, this is the case, and the strength for an infinite number of stress cycles is known as the endurance limit. For nonferrous metals and alloys, this is not true, and the strength used for repeated loading is the fatigue strength.

3-11 WEAR RESISTANCE

Wear is the unintentional displacement or removal of particles from the surface of a part due to rubbing. It can result in improper performance or fracture. Wear is dependent not only upon the characteristics of the metal of which the part is made, but upon the characteristics of the material that it is in contact with. There are several types of wear that require consideration in mechanical design. Abrasive wear, or erosion, is the removal of particles by the gouging action of the material that is rubbing against the part. Adhesive wear, or galling, is the intense interaction of two parts when forced together and there is motion between them. This interaction can result in large particles being torn from the surface of both parts, and the parts may even weld together.

Surface fatigue may be classed as wear. It is the fatigue failure of the metal immediately below the surface due to the frequent application of a load, as in the case of gear teeth. It takes the form of small thin flakes of metal being removed from the surface of the part. A corrosive environment can initiate or greatly accelerate wear, because of the effect it has on the surface of the part.

Because of the complex nature of wear, there is no single characteristic of a metal that can be used to evaluate the resistance to wear. Strength, toughness, and hardness are characteristics that are associated with wear resistance; hardness of the surface is the most important in the majority of cases. Other factors such as surface finish, lubrication, and the appropriate combination of metals for mating parts are also very important, and are discussed elsewhere in this text.

3-12 CORROSION RESISTANCE

Corrosion is the deterioration of a metal due to a reaction with its environment. Corrosion on a part can produce a film that protects the metal from further corrosion, or it can continue until the strength of the part is reduced to the point of failure. The reaction of a metal varies greatly with the environment. For example, a particular steel will corrode at a different rate in fresh water than in sea water, and at a different rate in dry air than in moist air.

Small variations in the constituents of a metal can greatly affect its resistance to a given environment. Characteristics of the environment, such as temperature and the degree to which the atmosphere is contaminated, have a considerable effect on corrosion. A corrosive environment can cause the fatigue strength to be much less than it would be under noncorrosive conditions. It can also cause the failure of a member that is stressed in static tension to a degree considerably less than the strength of the material.

3-13 MACHINABILITY

Machinability is the term used to specify the relative ease of machining a metal. There are several methods of determining machinability: power required, tool life, machining time, and surface quality. The more important factors are:

1. Hardness — The machinability rating varies roughly with hardness; generally, the harder a metal, the poorer its machinability rating.

2. Strength — The machinability rating varies with the strength of the metal. For steel there is a close relationship between machinability and the ratio of yield strength to tensile strength. Generally, the greater the strength the lower the machinability rating, but a metal can have too little strength to machine well.

3. Composition — Matter added intentionally or matter that is present unintentionally has a great effect on machinability. For example, the lead or sulfur that is added to steel considerably improves its machinability.

3-14 FORMABILITY

The formability of a metal is an indication of its suitability for parts that require severe forming, such as automobile fenders. Elongation is an important property in determining the formability of a metal, because of the great amount of stretching involved, at times over 40 percent. But high elongation is not enough. The metal must also have appreciable tensile strength to resist tearing. A metal with excellent forgeability may have poor formability due to low tensile strength or because it will not stand the cold working.

3-15 CASTABILITY

The term castability is not an indication of whether or not a metal can be cast, but an indication of how difficult it is to cast and the limitations imposed on the design of a part to make casting feasible. The most important factors determining the castability rating are:

1. The temperature at which the metal solidifies — the lower this temperature the easier it is to cast the metal.
2. The amount the metal shrinks after solidifying — the smaller this shrinkage the easier it is to make molds and the greater the flexibility in the design of the part.
3. The strength at the temperature just below that at which solidification occurs — the greater the strength the easier it is to produce sound castings and the greater the design flexibility.

3-16 FORGEABILITY

The forgeability of a metal is an indication of the difficulty involved in forging it. Forging is done when the metal is in the plastic condition. Generally, heating is necessary to produce this condition, but the temperature must not be permitted to go so high that it will burn the metal. Metals that do not become plastic, of course, cannot be forged. The difference between the temperature at which the metal becomes plastic and the temperature at which there is danger of burning is known as the plastic range. A large plastic range is desirable for forging, because of the greater time it allows to work the metal before reheating becomes necessary.

3-17 WELDABILITY

Though there are few metals that cannot be welded, a successful weld is more easily accomplished with some metals than it is with others. Weldability is the term used to indicate the ease with which a successful weld can be made. A weld involves not only making the joint, but the effect of the welding process on the elements being joined. Some of the important factors are as follows: the metal has a tendency to oxidize; the strength produced by cold working or heat-treating is largely lost near the weld; and the corrosion resistance near the weld may be reduced. Of course, these difficulties can be overcome or their effect greatly reduced by careful design and additional processes.

3-18 GRAY CAST IRON

Over three-fourths of all iron castings produced are of gray iron. Castings of gray iron are the cheapest castings that are produced and should be considered first whenever a casting is indicated. Because of the great effect that foundry procedures have, minimum tensile strength rather than composition has been adopted as the basis for the classification of gray iron castings. There are several systems of classification. The American Society for Testing and Materials (ASTM) system will be used here. It uses a number from 20 to 60 that indicates the minimum tensile strength in thousands of pounds; thus, class 30 has a strength of 30,000 psi. It should not be concluded that a higher number indicates a superior casting. For many applications the higher-strength castings are inferior, because of the changes in other characteristics that accompany the greater strength.

The low-strength castings can be produced without alloying elements, and the medium-strength castings require only alloying elements, but the high-strength castings require not only alloying elements but such close control that the low-cost method of melting in a cupola cannot be used.

Because of alloying elements, the close control in producing the molten metal, the higher molding costs to provide for greater shrinkage, and the greater number of rejections due to more rigid inspection, the high-strength castings cost considerably more than class 30. Classes 20 and 25 cost slightly more than class 30 because a smaller amount of cheap scrap iron can be used.

The modulus of elasticity varies with the class, being less for the low-strength castings. Fatigue resistance is very good for all gray iron, but it is slightly better for the low-strength irons. Wear resistance is excellent, and for the usual sliding members of machines is adequate as cast. The damping capacity is outstanding, especially for the low-strength castings. The machinability of carefully made gray iron castings is superior to most steel, but decreases with increasing strength. Gray iron is not appropriate where high impact resistance is required.

The characteristics of gray iron castings can be modified by heat-treatments. Annealing is effective in making a casting more machinable, but there is a considerable decrease in hardness and strength. Hardening is used to increase resistance to wear and abrasion, but seldom to increase tensile strength, as this can be more economically accomplished by using a higher class. Localized hardening processes, such as flame hardening, may be used where hardness is required in only small areas. Stress relief should be performed on castings that are to be machined, particularly if considerable accuracy is required. This heat-treatment is necessary to relieve the residual stresses that are present in any casting in the as-cast condition. The high-strength irons are more likely to have large residual stresses.

The minimum thickness for the sections of which a casting is composed varies with the class, the low-strength iron permitting much thinner sections. It is especially important to adhere strictly to the minimum section limitations if surfaces associated with these sections are to be machined, because thinner sections will probably produce "hard spots" that cannot be machined.

Typical applications are as follows:

Class 20: castings that must be made to close tolerance or with thin sections, or that require high machinability.
Class 30: general machinery, compressor bodies, engine blocks.
Class 50: heavy-duty machinery, diesel engine blocks.

Class 30 is most often specified, and it is representative of gray cast iron. Its properties are as follows: tensile strength = 30,000 psi; modulus of elasticity in tension = 14,000,000; compression strength = 110,000 psi; Brinell hardness = 200; shear strength = 42,000 psi; endurance limit = 14,000 psi; minimum section = 0.25 in.

3-19 WHITE CAST IRON

A white iron casting is very hard and thus has good wear resistance, but it is also very brittle. White iron is produced by adjusting the composition, and various alloying elements or combinations of elements may be added to produce a variety of characteristics and properties.

Because of its great hardness, white iron is not practical to machine and thus its use is rather limited. White iron is not as readily cast as gray; thus, fillets must be larger, changes in section should be made gradually, and sections less than 0.50 in. thick should not be used. The representative properties of unalloyed white cast iron are approximately as follows: Brinell hardness = 400; tensile strength = 30,000 psi; compressive strength = 200,000 psi; modulus of elasticity = 25,000,000; impact strength = one-third that of gray iron.

3-20 CHILLED CAST IRON

If gray iron is very rapidly cooled from the molten state, iron with characteristics of white iron rather than gray iron results. It is called chilled cast iron. Molds containing metal or graphite chillers are used to chill the molten iron at selected locations. Thus, hard wear-resistant surfaces are produced on a casting, the major portion of which has the characteristics and properties of gray iron.

Because of the differences in time of solidification in various parts of a chilled iron casting, and the difference in shrinkage characteristics of gray iron and chilled iron, very high residual stresses are often produced. These stresses can be relieved by suitable heat-treatment, which does not reduce the hardness of the chilled surfaces. A typical application of chilled cast iron is railroad car wheels, where a hard wear-resistant surface is desired where the wheel is in contact with the rails and brake shoes, but the impact resistance and damping characteristics of gray iron are desired for the remainder of the wheel.

3-21 MALLEABLE CAST IRON

Malleable iron castings are produced from white iron castings by a heat-treating process called malleableizing that takes several days. Though the cost is about twice as much as for gray iron castings, it is still less than for steel forgings, and the process can often be used in place of forgings in applications that require resistance to abuse and impact that cannot be provided by gray iron.

Its machinability is superior to that of free-cutting steel, but another method of providing the smooth surface and close tolerance can be used where quantities justify the tooling required. This method is to coin in a press the surfaces that would be machined. Another advantage is that sections as thin as 0.06 in. can be used.

Malleable iron castings are limited in size; the maximum weight is about 400 lb, the maximum length about 5 ft, and the maximum section thickness is 3 in. The preferred section thickness is about 1.50 in. The wear resistance and notch sensitivity of malleable cast iron is inferior to that of other cast irons. The two most common types, or grades, are 32510 (also known as grade II) and 35018 (also known as grade I). The major difference between them is that 35018 has greater ductility and slightly higher strength, and 32510 has better castability and slightly lower cost.

Grade 32510 is representative of malleable cast iron. Its properties are as follows: tensile strength = 50,000 psi; yield strength = 32,500 psi; elongation in 2 in. = 14 percent; modulus of elasticity = 25,000,000; and Brinell hardness = about 120. Typical applications are: gear cases, hand tools, and conveyor chains.

3-22 NODULAR CAST IRON

Nodular cast iron, also known as ductile cast iron, is a cast iron whose composition is similar to that of gray iron except for the addition of an element, or elements, to cause the structure to differ from gray iron and thus to have unique characteristics and properties. Additional alloying elements can be added and various heat-treatments can be employed to further modify the properties. The cost is slightly more than for malleable cast iron and slightly less than for cast steel.

Nodular iron castings are used successfully where resistance to impact is required and in applications where wear resistance is important. Its machinability is equal to, or better than, that of gray iron. The casting characteristics are the same as those of gray iron, except for greater shrinkage during solidification, which can easily be provided for by properly placed risers.

Types of nodular iron are identified by a system involving three numbers such as 60-45-10; the first indicates the minimum tensile strength in 1000 psi, the second, the yield strength in 1000 psi, and the third, the percentage of elongation in 2 in. The three types most often used are 60-45-10, 60-40-15, and 60-45-15. The modulus of elasticity for all types is about 23 million and the compressive strength is approximately equal to the tensile strength. The damping capacity is not quite as good as for gray iron, but it is twice as good as for cast steel. Applications include automobile crankshafts, paper machinery press rolls, and generator shafts, some of which weigh over 5 tons.

3-23 CORROSION AND HIGH-TEMPERATURE SERVICE

Gray cast iron rusts, but at a much slower rate than steel. Often an adherent coating of rust forms that serves to protect the part from further attack. However, there are a number of factors that affect the formation of this protective coating and its tendency to adhere to the surface, such as composition of the metal and turbulence in the surrounding gas or liquid

Fig. 3-2 A large turbine-generator shaft. (*Bethlehem Steel Corp.*)

which scrubs off the coating. The resistance to various chemicals is dependent not only on the type of chemical but on its concentration and temperature. The performance of gray cast iron carrying natural water is very good, but its performance when buried depends on the soil. If the soil is well drained there is generally an acceptable corrosion rate. The corrosion rate in sea water is relatively low. Gray cast iron is often used as a container for molten metal.

One of the unique characteristics of gray cast iron is that it grows when subjected to high temperature. This growth is not to be confused with thermal expansion. Another adverse characteristic is the formation of scale on the surface, which can be either adherent or flaking. If the temperature is less than 750°F these characteristics are not a problem. Strength is also not appreciably affected by temperatures less than 750°F.

The characteristics of malleable and nodular iron are similar to those of gray iron except for the superior performance of nodular iron at high temperature.

Many special types of cast iron are produced by the use of alloying elements, which have superior performance in a particular situation involving corrosion or heat. Cost is only one of the advantages that may be sacrificed by resorting to these special types.

3-24 STEEL

With about 10,000 different types and grades of steel available, it is obvious that this discussion will have to be limited to basic principles, with detailed information given for only a few representative types. With this great number of steels from which to choose, it may be concluded that little of any one grade would be found in use, but this is not the case. In alloy steels, fewer than 25 grades account for over half of the production.

The shaft shown in Fig. 3-2 is an excellent example of a part the requirements for which can only be satisfied by a high-quality steel. It will be used in a 620,000-kw nuclear-powered turbine generator. It was machined from a 202-ton forging. The diameter is 65.25 in., and the length is 46 ft 11.75 in.

Wrought steels are produced by rolling in a steel mill and are usually divided into two major categories: plain carbon steels and alloy steels. The generally recognized method of specifying a steel is based on the classifications of the Society of Automotive Engineers (SAE). In this system, the first two digits indicate the type of alloy and the last two (sometimes three) indicate the carbon content in hundredths of 1 percent. The various types and their basic numbers are shown in Table 3-2. The letter B following the first two digits indicates a boron-intensified steel, and an L indicates that lead has been added.

TABLE 3-2 SAE-AISI Steel Designations

FIRST TWO DIGITS	TYPE
10	Plain carbon
11	Free-cutting
13	Manganese
25	Nickel
31, 33	Nickel-chromium
40	Molybdenum
41	Chromium-molybdenum
43, 47, 86, 87, 93, 98	Nickel-chromium-molybdenum
46, 48	Nickel-molybdenum
50, 51, 52	Chromium
61	Chromium-vanadium
92	Silicon-manganese
94	Manganese-nickel-chromium-molybdenum

The American Iron and Steel Institute (AISI) classification uses the same numbers as the SAE system and adds a prefix to indicate the method of steel manufacture. The prefixes are as follows:

B — Acid Bessemer carbon steel
C — Basic open-hearth carbon steel
E — Electric-furnace steel

No prefix indicates open-hearth alloy steel.

For example, C1008 is a plain carbon steel, made by the open-hearth method, that has a nominal carbon content of 0.08 percent; 3140 is a nickel-chromium alloy steel, with a carbon content of 0.40 percent, and is produced by the open-hearth method.

Plain carbon steels account for about 80 percent of the steel produced. They are cheaper than alloy steels and should be considered whenever steel is indicated. They are further broken down into: low-carbon steels, medium-carbon steels, and high-carbon steels.

3-25 LOW-CARBON STEELS

Low-carbon steels are those which contain less than 0.30 percent carbon and are generally not hardened. Those steels with a carbon content of less than 0.15 percent have relatively low strength, but cold working will increase the strength and decrease the ductility. They do not machine well and should not be used for parts involving threading, broaching, or where a smooth machined finish is required. They are used where good drawing qualities are required, as in automobile fenders. They are also used in applications involving cold heading, such as rivets, and where welding is required.

Those with a carbon content of between 0.15 and 0.30 percent have greater strength and are often used for forging and case hardening — 1018 is used for thin sections and 1024 for heavier sections. 1020 is often used for cold heading. All of the steels in this carbon range are suitable for welding, and they machine better than those of less than 0.15 percent carbon. An important group of steels are those known as "free-cutting," some of which have a carbon content of less than 0.30 percent. These will be discussed in Sec. 3-30.

3-26 MEDIUM-CARBON STEELS

Medium-carbon steels are those that have a carbon content of 0.30 percent to about 0.60 percent. They have wide application where strength and hardness requirements cannot be met by low-carbon steels. These steels may be heat-treated to produce a wide range of properties and are the most versatile of the plain carbon steels.

They can be used for forging — 1030 for small parts, 1036 for more critical parts, and 1048 for larger forgings such as crankshafts. Many parts are machined from bar stock, both with and without subsequent heat-treatment. 1035 is often used for bolts that are made by cold heading.

These steels are suitable for welding, but greater care is required than when welding low-carbon steel.

3-27 HIGH-CARBON STEELS

High-carbon steels are greatly inferior to medium-carbon steels in machinability, weldability, and formability. They are also more brittle when heat-treated. They are not used as widely as the carbon steels of lower carbon, but they are used in a number of important applications where high strength, hardness, or good wearing qualities are the primary requirements.

Manufacturers of farm implements make use of these steels. Because of their greater strength, many parts are made without heat-treatment, and thus adequate strength is achieved at a minimum cost. Other parts, such as disks for plows and mower knives, are hardened to provide a wear-resistant cutting edge. Hand tools, such as wrenches and hammers, are made from these steels. They are also used for springs (discussed in Chap. 17) and metal-cutting tools (discussed in Sec. 3-34). It must be remembered that these steels are shallow-hardening when compared with alloy steels.

3-28 HOT-FINISHED STEEL

Hot-finished steel is that which is rolled into bars, plates, and structural shapes while hot. Generally, the temperature when the rolling is completed is over 1200°F. Low-carbon and medium-carbon steels are often used as rolled, but high-carbon steel is usually annealed before the manufacturing processes are performed, and then it is heat-treated.

Hot-finished low-carbon steel has the lowest cost of any steel. The greatest amount is used in buildings, ships, bridges, and railroad cars. It is often produced to a quality grade that imposes certain requirements important to the intended use, such as: bridges and buildings (ASTM A7), locomotives and cars (ASTM A113), and boilers and other pressure vessels (ASTM A299).

Hot-finished steel is usually the plain carbon type. An important exception is the high-strength low-alloy steels, which have both greater strength and greater corrosion resistance than the ordinary structural steel. This improvement is accomplished by adding moderate amounts of one or more alloying elements other than carbon.

These steels are produced to property requirements rather than to a chemical composition specification. The properties are, generally: yield strength = 50,000 psi; tensile strength = 70,000 psi or slightly greater; and elongation = about 20 percent for a thickness of less than 0.50 in. (for a greater thickness the strength is slightly less).

These steels are readily welded with no special procedures required, and the corrosion resistance is several times that of ordinary structural steel. They are used principally in mobile equipment but are also used in other applications where reduced weight or greater corrosion resistance justifies the cost, which is about 50 percent greater than for ordinary structural steel.

3-29 COLD-FINISHED STEEL

All cold-finished steel begins as hot-rolled. There are two methods of cold finishing: one is to remove material by turning or grinding to produce a good finish and the proper size; the other is to produce the shape of the cross section (which can be other than circular), the proper size, and and the finish by cold rolling or cold drawing.

Bars finished by turning or grinding range from 0.50 to 9 in. in diameter. Those produced by cold drawing are generally less than 4 in. in diameter or maximum dimension if the cross section is other than circular.

The properties of the steel are not altered by turning or grinding, but they are by cold rolling or cold drawing. Cold drawing is especially effective in increasing the yield strength; only 8 percent reduction produces an increase of 60 percent. This is accompanied by a 30 percent decrease in elongation and an increase in tensile strength of less than 15 percent. Machinability is also improved by cold drawing.

Carbon steels with a carbon content of over 0.55 percent, or alloy steels with more than 0.40 percent carbon, are seldom cold-drawn, as they must be annealed before drawing, and there is little advantage gained to offset the cost of annealing.

Sheets thinner than 0.0598 in. are cold-rolled, because the metal cools too rapidly for hot rolling to be practical. The range of thickness for cold-rolled sheet is from 0.1196 to about 0.0149 in. Cold-rolled sheet is also produced to various quality grades for particular uses, such as deep drawing. Cold-finished bars cost 2 to 3 cents/lb more than hot-finished bars.

3-30 FREE-CUTTING STEELS

Free-cutting steels are those to which elements such as sulfur, lead, or phosphorus have been added for the purpose of improving the machinability. These steels cost slightly more than those which do not have these additions and should, therefore, not be used unless the higher cost is offset by the lower machining costs. The point at which such steels should be considered is when about 20 percent of the metal becomes chips in machining the part.

Steels to which sulfur has been added should not be used for parts involving cold heading, but steels to which lead has been added may be used. Neither is preferred for welding.

Both sulfur and lead may be added. Such a steel is appropriate for parts where more than half of the metal becomes chips. Adding sulfur or lead in amounts necessary to improve machinability has little, if any, effect on the other characteristics or properties of either plain carbon or alloy steels.

3-31 PROPERTIES OF PLAIN CARBON STEELS

The properties of some of the more frequently used plain carbon steels are given in Table 3-3. From an examination of this table it is seen that slightly greater strength can be obtained by using the cold-drawn condition than by heat-treating, when the tempering temperature is 1300°F or more. The effect of the tempering temperature on a representative medium-carbon steel is shown in Table 3-4. The effect of size on the properties of a representative medium-carbon steel, water-quenched and tempered at 800°F, is shown in Table 3-5. The yield strength and hardness vary with both the tempering temperature and the size, but the elongation varies only with the tempering temperature.

TABLE 3-3   Properties of Carbon Steels

TYPE OF STEEL	TYPE OF PROCESSING	TENSILE STRENGTH, PSI	YIELD STRENGTH, PSI	ELONGATION, % IN 2 IN.	BRINELL HARDNESS NO.
1008	Hot-finished	44,000	24,500	30	86
	Cold-drawn	49,000	41,500	20	95
1015	Hot-finished	50,000	27,000	28	101
	Cold-drawn	56,000	47,000	18	111
1030	Hot-finished	68,000	37,500	20	137
	Cold-drawn	76,000	64,000	12	149
	HTT*(400°F)	122,000	93,000	18	495
	HTT(1300°F)	75,000	58,000	33	179
1050	Hot-finished	90,000	49,500	15	179
	Cold-drawn	100,000	84,000	10	197
	HTT(400°F)	143,000	108,000	10	321
	HTT(1300°F)	96,000	61,000	30	192
1095	Hot-finished	120,000	66,000	10	248
	HTT(400°F)	188,000	120,000	10	401
	HTT(1300°F)	109,000	74,000	26	229
1112	Hot-finished	56,000	33,500	25	121
	Cold-drawn	78,000	60,000	10	167
1144	Hot-finished	97,000	53,000	15	197
	HTT(400°F)	128,000	91,000	17	277
	HTT(1300°F)	97,000	68,000	24	201

*HTT is heat-treated and tempered at the temperature shown.

For all plain carbon and alloy steels, the modulus of elasticity is between 29 and 30 million, the density is 0.28 lb/cu in., the shear yield strength may be taken as 55 percent of the tension yield, and the bearing yield as 75 percent of the bearing ultimate.

3-32   ALLOY STEELS

Alloy steels are those whose properties are greatly modified by the presence of alloying elements in amounts substantially greater than are

TABLE 3-4   Effect of Tempering

TEMPERING TEMPERATURE, °F	TENSILE STRENGTH, PSI
600	106,000
800	101,000
1000	94,000
1200	86,000

TABLE 3-5 Effect of Size

DIAMETER, IN.	TENSILE STRENGTH, PSI
1	140,000
2	115,000
3	108,000
4	105,000

found in plain carbon steels. Although alloy steels are more expensive than carbon steels, the added expense is justified in instances where their selection has been made upon careful evaluation of the advantages, some of which may be: greater resistance to wear, fatigue, creep, and corrosion; greater toughness; or greater tensile strength with little sacrifice of ductility. In many instances, the greater strength makes it possible to use less of the more expensive steel, thus partially offsetting its higher cost.

One of the major reasons for alloying steel is to increase the depth of hardening. This is illustrated as follows: a 5-in.-diameter bar of 4140 can be hardened through to a tensile strength over twice as great as a 1040 bar. Putting it another way, a 4140 bar can be hardened through to Rockwell C45 if it is 3.5 in. in diameter, but a 1040 bar can be hardened through to this degree only if it is smaller than 0.65 in. in diameter.

The effect of alloying elements can be summarized as follows:

Aluminum Increases hardness of nitriding steels.

Boron Increases hardness.

Chromium Increases tensile strength, hardenability, and abrasion resistance; reduces ductility.

Copper Improves resistance to atmospheric corrosion.

Manganese Improves tensile and impact strength, abrasion resistance, and hardenability.

Molybdenum Increases tensile strength and hardenability.

Nickel Increases tensile strength, fatigue resistance, toughness, and corrosion resistance.

Phosphorus Improves strength, machinability, and corrosion resistance; reduces toughness.

Silicon Improves hardenability, strength, and electrical properties.

Sulfur Improves machinability and decreases toughness.

Vanadium Improves tensile, fatigue, and impact strength; improves resistance to creep and abrasion.

Alloy steels are classified as low-carbon, medium-carbon, and high-carbon. The low-carbon steels are used mainly for parts that are to be case

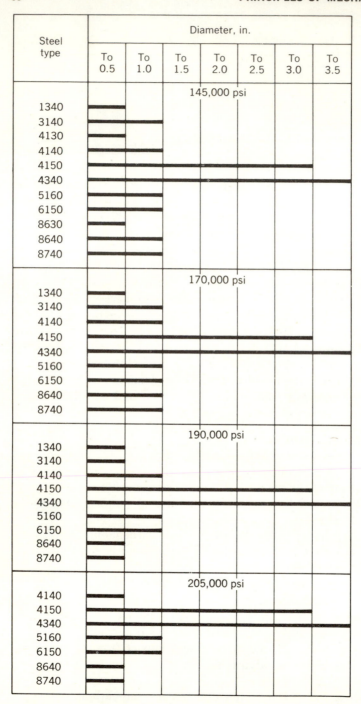

Fig. 3-3 The variation of strength with alloy and size.

hardened. Case hardening will be discussed in Sec. 4-10. Alloy steels have considerably greater core strength than do plain carbon steels of the same carbon content. The medium-carbon alloy steels are used much the same as the plain carbon steels of medium-carbon content. They are preferred for parts that are highly stressed, subjected to impact or fatigue, and for large parts that could not be hardened completely if plain carbon steel were used. They are also preferred where weight is an important factor. The high-carbon steels are used primarily for springs and parts where wear resistance is a major consideration.

A list of alloy steels has been compiled from those that are often used and are readily available. They, of course, have different heat-treating characteristics. Figure 3-3 gives the ultimate tensile strength that can be achieved for various sizes of these selected steels. At the top of the figure, sizes are given as 0.5 in. in diameter, 1.0 in. in diameter, and so on. The figure is divided into four sections; the first is titled 145,000 psi. The horizontal lines in this section indicate the size that can be completely heat-treated to a strength of 145,000 psi. Thus, 1340 can be heat-treated through to 145,000 psi if not more than 0.5 in. in diameter, 3140 can be heat-treated to 1.0 in. in diameter, and 8630 can be heat-treated to 0.5 in. in diameter. In the next section, which is titled 170,000 psi, 8630 does not appear. This means that a bar of 8630 that is 0.5 in. in diameter cannot be heat-treated through to a strength of 170,000 psi. It should be noted that of the initial eleven steels, only seven can be heat-treated through to a strength of 205,000 psi when 0.5 in. in diameter, and that only two can be heat-treated to any of the strengths indicated when the bar is over 1.0 in. in diameter.

If the form of the steel is flat, rather than a circular cross section, the information in Fig. 3-3 is applicable if the equivalent thickness given in Table 3-6 is substituted for the diameter. Thus, 1340 can be heat-treated through to 145,000 psi if not more than 0.3 in. thick, or 3140 can be heat-treated through to 145,000 psi if not more than 0.6 in. thick.

TABLE 3-6 Equivalent Thickness

DIAMETER, IN.	THICKNESS, IN.
0.5	0.3
1.0	0.6
1.5	1.0
2.0	1.3
2.5	1.6
3.0	2.0
3.5	2.3

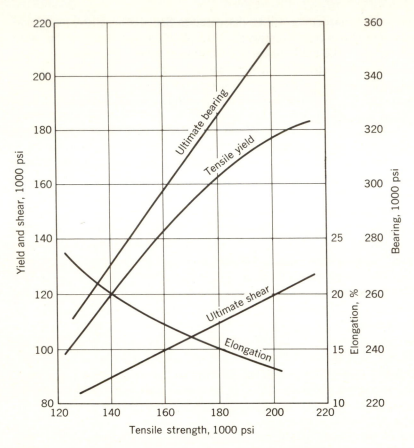

Fig. 3-4 The relation of other properties to tensile strength.

The relation of other properties to the tensile strength for medium-carbon heat-treated alloy steels is given in Fig. 3-4. The approximate Brinell hardness is found by dividing the tensile strength by 500.

The H steels are a series that are designated by AISI and SAE numbers, to which an H is added as a suffix. They are similar in composition, and very similar in properties, to their equivalents without the H. The difference is that they are produced to meet specified hardenability requirements rather than to a specified composition.

In an effort to produce greater strength with adequate toughness and ductility, 4330 has been modified and has a tensile strength of 250,000 psi; 4340 is heat-treated in a nonstandard manner to produce a tensile strength of over 270,000 psi. Crucible Steel Co. has developed a steel called Hi-Tuf that has a tensile strength of 230,000 psi, but in addition has much better impact resistance than those just mentioned.

Another of the ultra-high-strength steels is the maraging type. This type is a high-nickel alloy that can produce a tensile strength of over 300,000 psi by a unique method of heat-treating that does not involve quenching.

3-33 HIGH-ALLOY STEELS

The most common high-alloy steels are those known as stainless steels. They contain from 12 to 30 percent chromium and some have, in addition, up to 22 percent nickel. These elements are responsible for the resistance to corrosion that makes stainless steel desirable for many applications. This resistance to corrosion is due to an oxide film that forms on the surface and prevents further oxidation. Passivation is a chemical treatment that is sometimes used to promote the formation of this film.

Stainless steels may be classified into three basic groups: those that can be heat-treated to improve their properties, known as martensitic; those that cannot be heat-treated, known as ferritic; and those that cannot be heat-treated, but can be strengthened appreciably by cold working, known as austenitic. All of these steels that are standard are designated by AISI-type numbers consisting of three digits.

The steels in the martensitic group have excellent resistance to mild corrosives such as the atmosphere, fresh water, and gasoline. Type 420 (the original stainless steel) can be heat-treated to a tensile strength of 250,000 psi. Forging and machining can be accomplished without difficulty if attention is given to the procedures used. 420 is seldom welded and is never used in the annealed condition.

Type 410 is much more frequently used and can be taken as representative of this group. Its properties when annealed and when hardened to its maximum strength are given in Table 3-7. Intermediate properties can be obtained by varying the heat-treatment.

Type 410 can be readily hot-worked or cold-worked to a degree, and it machines about as medium-carbon alloy steel does. It can be welded by the usual methods and need not be hardened to develop its maximum

TABLE 3-7 Properties of Type 410

	ANNEALED	HARDENED
Tensile strength, psi	75,000	195,000
Yield strength, psi	40,000	150,000
Elongation (% in 2 in.)	35	15
Brinell hardness	155	390
Endurance limit, psi	40,000	

TABLE 3-8   Properties of Type 430

Tensile strength, psi	75,000
Yield strength, psi	45,000
Elongation (% in 2 in.)	30
Brinell hardness	155
Endurance limit, psi	40,000

corrosion resistance. This steel is appropriate for elevated-temperature applications to 1000°F where corrosion resistance and moderate strength are required. The creep strength (load for 1 percent elongation in 10,000 hr) at this temperature is 9200 psi.

Type 431 has the best corrosion resistance of this group. Type 440 can be hardened to 653 Brinell.

The steels in the ferritic group are more resistant to corrosion than those in the martensitic group, and the nonhardening tendency is an advantage when hot working or welding is involved. Type 430 is representative of this group and is the most often specified. Its properties are given in Table 3-8. This type can be readily hot-worked and cold-worked; even deep drawing can be accomplished with care. The machining characteristics are similar to those of type 410; the creep strength (load for 1 percent elongation in 10,000 hr) at 1000°F is 8500 psi. Type 446 has the highest corrosion resistance of the ferritic group.

The steels in the austenitic group are the most resistant to corrosion of any of the stainless steels and have exceptional toughness and ductility. They are more difficult to machine than the other stainless steels. They are also noted for their strength and oxidation resistance at high temperatures.

Type 302 is representative of this group; its properties when annealed and in the full-hard condition are given in Table 3-9. This type may be hot-worked, and its cold-working characteristics are superior to those of 410 or 430. It can be welded easily by standard methods, but the weld area

TABLE 3-9   Properties of Type 302

	ANNEALED	FULL-HARD
Tensile strength, psi	85,000	110,000
Yield strength, psi	35,000	75,000
Elongation (% in 2 in.)	60	35
Brinell hardness	150	240
Endurance limit, psi	34,000	

must be annealed after welding. It can be machined, but is less desirable in this respect than 410 or 430. The creep strength (load for 1 percent elongation in 10,000 hr) at 1000°F is 17,000 psi.

Type 303 is the machining type for this group. Type 316 is the most corrosion-resistant of this group and has superior heat resistance. Its creep strengths (load for 1 percent elongation in 10,000 hr) for several temperatures are given in Table 3-10.

3-34 TOOL STEELS

Tool steels are steels of the highest quality, primarily intended to be used to make the cutting tools, forging dies, stamping dies, etc., that are required in the manufacture of parts from other materials. To serve such diverse purposes as maintaining a sharp cutting edge at high temperature, or maintaining a sharp cutting edge and resisting impact, or resisting impact at high temperature, a number of different types is necessary.

Tool steels cost much more than other steels. The alloying elements account for some of this difference, but the care and precision required to produce them is responsible for most of it. The assurance of the higher quality justifies the cost when the great amount of time involved in making the tools is considered.

Tool steels of the plain carbon type that are hardened by quenching in water are the least expensive. The carbon content is approximately 0.70 to 1.40 percent. They are probably the most widely used, even though they are inferior to the alloy types in hardenability and in wear resistance, and have a greater tendency to deform when heat-treated. The surface can be hardened to Rockwell C65. Typical uses are files, drills, and blanking dies.

The shock-resisting steels maintain resistance to impact when hardened to Rockwell C60, and are used for tools such as chisels and hammers. Some of the hot-work group resist softening at temperatures of 1000°F and at the same time have a tensile strength approaching 300,000 psi. They

TABLE 3-10 Creep Strength of Type 316

TEMPERATURE, °F	CREEP STRENGTH, PSI
1000	25,000
1100	18,000
1200	13,000
1300	8,000

are used for hot dies such as those required for forging. Those known as high-speed steels have the ability to maintain a keen cutting edge at the temperatures that result from operating the cutting tools at a much higher speed than can be used with other steels that are used for cutting tools.

The use of tool steels is not limited just to tools. They are used in other applications where a particular characteristic or property is required and the cost is justified. This usually involves a small part or a very special situation, for the cost of tool steel can amount to several dollars per pound.

3-35 CAST STEELS

A relatively small percentage of the steel produced is used for steel castings. But much of the steel that is used for casting is for rather important applications, where the strength and ductility of steel and the advantages of the casting process are required for a wholly successful design. Steel castings are more difficult to produce than are iron castings, in part because of the higher pouring temperature, which means greater shrinkage and internal stresses.

Cast steel can be either plain carbon or alloy, and the plain carbon can be further broken down into low-carbon, medium-carbon, and high-carbon. Low-carbon steel castings are those which have a carbon content of less than 0.20 percent. If a casting is carefully designed to minimize the internal stresses, it may be used as cast, but otherwise it should be annealed. Railroads make the most extensive use of low-carbon steel castings.

Medium-carbon castings have a carbon content between 0.20 and 0.50 percent. Well over half the steel castings are of the medium-carbon type. Whenever possible, they should be heat-treated by liquid quenching and tempering, because of the improvement in properties. As an example, for a casting of 0.30 percent carbon content the tensile strength is increased from 80,000 to 122,000 psi by quenching and tempering at 700°F.

The modulus of elasticity is the same as for wrought steel. The ultimate compression strength, the compression yield strength, and the modulus of elasticity in compression are essentially the same as for tension. The endurance limit for low-carbon and medium-carbon castings is usually taken as 40 percent of the tensile strength. The fluidity of steel is less than for other metals; thus, a minimum section of 0.25 in. is recommended for small castings. As the size of the casting increases, so does the minimum section thickness, and for a large casting it may be over 1 in. As the size of the section increases, the strength decreases. There may be a decrease of as much as 5000 psi between a section of 1 in. and a section of 6 in. It is as easy to get good welds between two castings, or between a casting and wrought steel, as it is between two pieces of wrought steel.

The principal use of high-carbon steel castings is in applications that require great hardness or abrasion resistance.

Low-alloy cast steels have characteristics and properties very similar to those of wrought steels of the same alloy and carbon content, and they respond similarly to heat-treatment. Many steel castings are produced in accordance with specified properties, rather than with specified composition, in the same manner as iron castings.

High-alloy cast steels are used in applications where resistance to corrosion or heat is required to a greater extent than is provided by the low-alloy steels. Some of these steels are used in applications where both corrosion and high temperature are involved. The high-alloy designations consist basically of two letters: those beginning with C are intended for corrosion-resistant applications below 1200°F, and those beginning with H are intended for use above 1200°F, some of which are used at 2000°F.

3-36 ALUMINUM

It is common practice to use the word aluminum not only for commercially pure aluminum, but for the numerous aluminum alloys. The cost of aluminum is several times the cost of low-carbon steel per pound. It weighs 0.098 lb/cu in., which is one-third as much as steel, and the strength of some high-strength alloys is greater than for low-carbon steel.

Pure aluminum has good resistance to atmospheric corrosion and to many chemicals, but the corrosion resistance of the high-strength alloys is inferior. One method of dealing with this situation is to roll a thin layer of pure aluminum or a resistant alloy into each side of the sheet. Such a composite is known as clad aluminum. Another way of dealing with the problem is to anodize the part. This treatment imparts a hard, corrosion-resistant coating to the aluminum. This treatment is not limited to sheet. Aluminum can be plated and, of course, painted.

Relative to the corrosion resistance of aluminum, several items must be mentioned. Non-heat-treatable alloys have greater corrosion resistance than do those that can be heat-treated. Aluminum is not resistant to strong alkalis. It is also subject to corrosion when in contact with a dissimilar metal, but this can be prevented in most cases by painting the facing surfaces with zinc chromate primer. Some of the high-strength alloys are susceptible to intergranular corrosion if improperly heat-treated.

Aluminum is noted for its exceptional electrical conductivity, thermal conductivity, and reflectivity. The electrical conductivity is about 60 percent that of copper, but because of the difference in density, an aluminum conductor of resistance equal to that of a copper conductor weighs about one-half as much. This largely accounts for the fact that most power transmission lines in the United States employ aluminum.

The thermal conductivity of aluminum is over 50 percent that of copper; cast iron has only 10 percent. This accounts for its selection for use in refrigerators and cooking utensils. The performance of aluminum is satisfactory at temperatures as low as 400°F below zero. At low temperatures there is a slight increase in strength and a considerable decrease in elongation. At elevated temperatures aluminum rapidly loses strength. For example, one of the widely used high-strength alloys, after 10,000 hours at 500°F, loses 80 percent of its tensile strength, and its elongation is doubled. The coefficient of expansion of aluminum is 0.000013 in./(in.) (°F), which is about 3 times that of steel. This difference becomes important when close tolerances are involved in a mechanism that has both steel and aluminum parts.

Though soft aluminum is rather difficult to machine satisfactorily, the harder grades machine very well, and if attention is given to controlling the temperature, close tolerances can be held. Any strength developed by cold working or heat-treatment is lost in the area heated by welding, though heat-treatment after welding restores the strength of heat-treatable alloys. The fact that aluminum does not produce sparks when struck is important in some applications, and aluminum is therefore widely used in the manufacture of explosives.

Aluminum is classed as wrought or cast. All wrought aluminum is identified by a four-digit number, and castings are identified by a two-digit or three-digit number. Sometimes a letter is a part of the casting designation. Some wrought aluminum is strengthened by strain hardening (work hardening). The following system is used to indicate the magnitude of hardness, or temper, and any other treatment that may affect the strength.

H1 indicates strain-hardened only. The second digit will specify the temper in accordance with the following: 8 is full-hard, 4 is half-hard, 2 is quarter-hard, and so on. Thus, 1100-H18 means wrought alloy 1100 strain-hardened to the full-hard condition. A third digit may be used to identify special properties.

H2 indicates strain-hardened and partially annealed. In this case, the work hardening produces greater hardness than is desired, and the partial annealing brings the temper to that indicated by the second digit, which has the same meaning as before. For example, 5456-H24 indicates wrought alloy 5456, strain-hardened and partially annealed to the half-hard condition.

H3 indicates strain-hardened and stabilized. This stabilizing treatment is used only with alloys containing magnesium, and it is required to prevent a gradual softening at room temperature. The second digit has the same meaning as before.

Some aluminum can be strengthened by heat-treating. Heat-treatment is indicated by adding a T to the alloy number. The number following the T indicates the kind of heat-treatment. A second digit may be added to indicate a minor variation in the heat-treatment that causes slightly different properties; it is not an indication of temper. The alloy designation may be followed by just an F or an O. An F means that the product is supplied in the as-fabricated condition. An O is used to indicate wrought products that are annealed to the softest condition.

3-37 WROUGHT ALUMINUM

Wrought aluminum can be strengthened by either cold working or heat-treating. An important advantage of the heat-treatable alloys is that severe forming may be done before heat-treating when the metal is soft. Any aluminum that can be heat-treated should be in the heat-treated condition when the part is ready for assembly. Heat-treatable alloys should be purchased in the heat-treated condition except where forming is to be performed.

The following extensively used alloys have been selected as representative of wrought aluminum. The properties of each are given in Table 3-11. The values in the table are average. The fatigue strength is for 500,000,000 cycles. These approximate values apply to all aluminum: modulus of elasticity = 10,500,000; modulus of rigidity = 4,000,000; shear yield strength = 0.55 times the tension yield. Comparisons are relative to aluminum unless stated otherwise.

TABLE 3-11 Properties of Aluminum Alloys

	STRENGTH IN 1000 PSI				
ALLOY	TENSION, ULTIMATE	TENSION YIELD, 0.2% OFFSET	SHEAR, ULTIMATE	FATIGUE STRENGTH	ELONGATION, % IN 2 IN.
2024-T3	70	50	41	20	18
6061-T6	45	40	30	14	17
7075-T6	83	73	48	23	11
3003-0	16	6	11	7	40
3003-H14	22	21	14	9	16
3003-H18	29	27	16	10	10
356-T6	33	24	26	8.5	3.5
220-T4	48	26	34	8	16
108F	21	14	17	11	25
380F	48	24	31	21	3

2024 is a heat-treatable alloy available in many forms, including sheet, plate, bar, rod, extrusions, and forgings. The nominal tensile strength is 64,000 psi when in the T4 condition. For sheet, the strength is slightly less, and for plate over ½ in. thick, the strength decreases with increasing thickness. Rolled and cold-finished rod and bar aluminum has a tensile strength slightly less than this nominal strength, and extrusions are even weaker. The corrosion resistance and cold-working characteristics are inferior, and it should not be brazed or gas welded, though arc and resistance welding may be accomplished. This alloy is used in aircraft structures, in truck wheels and bodies, and as screw machine stock. Clad 2024 has greatly increased resistance to corrosion and slightly lower strength.

6061 is a heat-treatable alloy, available in many forms, that may be readily cold-worked, welded, and brazed. It has very good corrosion resistance. It is used in structures, truck frames, railroad cars, and furniture.

7075 is a heat-treatable alloy of great strength and hardness in the T6 condition; it is also available in the O condition. It has inferior corrosion resistance bare, but it is available clad. It cannot be brazed and should be welded only by resistance welding. It is used mainly in aircraft structures.

3003 is a non-heat-treatable alloy that is available in many forms. The significance of cold working is evident when the strengths given in Table 3-11 for the various tempers are compared. In the clad form the strength is slightly less. The maximum thickness available for the different tempers decreases from 3 in. for the O condition to 0.128 in. for the H18 condition. This is because the strength of strain-hardened alloys depends on the amount of cold work.

This alloy in all tempers has good corrosion resistance. It is readily brazed or welded, and is easily cold-worked in the softer tempers. It is used for cooking utensils, chemical equipment, and architectural applications.

3-38 ALUMINUM CASTINGS

Aluminum castings are usually produced by sand casting, permanent-mold casting, or die casting. Heat-treating may be used to produce greater strength with some alloys. The minimum section is dependent, to a considerable extent, on the casting process: 0.12 in. for lengths less than 3 in. is easily achieved in sand castings, and 0.05 in. for small die castings.

Alloy 356 is frequently used and may be taken as representative. It is a heat-treatable alloy, with excellent casting characteristics, and is readily welded. It has good corrosion resistance and fair machining qualities. It is used for either sand or permanent-mold castings. The permanent-mold process produces several thousand psi greater strength, because of the more rapid cooling. The properties given in Table 3-11 are for the sand casting.

Typical uses include automotive gear cases and housings, pump bodies, water-cooled cylinder blocks, and aircraft fittings.

Alloy 220 is a heat-treatable alloy used for sand casting. It has very high strength and excellent corrosion resistance and machinability, but it requires special foundry practices. It is used in applications requiring both strength and shock resistance.

Alloy 108 is a general-purpose sand-casting alloy. It cannot be heat-treated, has good casting characteristics, and may be welded. It has poor corrosion resistance and should not be brazed. It may be used where pressure tightness is required. Alloy A108 is a permanent-mold alloy that is very similar to alloy 108 but has about 30 percent greater strength.

Alloy 380 is used for die casting. It is non-heat-treatable, has excellent casting characteristics, has good corrosion resistance, and machines well. It is used for the majority of aluminum die castings.

3-39 COPPER ALLOYS

The characteristics of copper and its alloys of greatest significance are electrical and thermal conductivity, corrosion and fatigue resistance, and the ease with which manufacturing processes may be performed. Commercially pure wrought copper is used to some extent, mainly for electrical conductors, but for most applications an alloy is preferred. There are hundreds of different copper alloys. An important use is as plain bearings. This use and the appropriate alloys will be discussed in Chap. 12.

Generally, copper alloys are suitable for use at temperatures as low as −300°F, but they rapidly lose strength at temperatures higher than 550°F. Copper does not produce satisfactory castings, and plain copper-zinc alloys are also seldom cast. Because of the effect on the molds of the high temperatures required, copper alloys are not used to any extent for die castings or permanent-mold castings.

Wrought copper and copper alloys are available in various degrees of hardness or tempers that are dependent on the amount of cold working that is performed after the last annealing. These tempers are: spring, hard, three-quarter-hard, half-hard, and quarter-hard.

Brass is basically an alloy of copper and zinc, while bronze is an alloy containing one or more elements other than zinc. The Copper Development Association has proposed a system of designating various wrought-copper alloys by three-digit numbers in place of the confusing trade names. This system will be used here to specify a particular alloy.

3-40 BRASS

Copper-zinc alloys are the least expensive of the copper alloys, especially those with a large zinc content. They cannot be heat-treated.

Those with less than 15 percent zinc are suitable for extreme cold working, but they are difficult to machine. With a zinc content over 35 percent, cold working becomes difficult, but hot working is easily performed. A zinc content of over 40 percent is seldom used.

Type 260 contains 30 percent zinc. The forms available include: strip, bar, rod, sheet, and tube. It is widely used. Some applications are: radiator cores, tanks, electrical sockets, and ammunition. Properties for two tempers are given in Table 3-12. The yield strength given in the table is based on a 0.5 percent extension under load. The density of copper alloys is about 0.31 lb/cu in., and the coefficient of expansion is 0.000011 in./(in.)(°F).

Type 360 is intended for parts made by machining, and Type 377 for forgings. Their properties are also given in Table 3-12.

3-41 BRONZE

Bronzes can be classified as those that can be heat-treated and those that cannot. Both heat-treatable and non-heat-treatable bronzes are produced as wrought products and castings.

Phosphor bronzes are non-heat-treatable, but they have high strength

TABLE 3-12 Properties of Copper Alloys

| TYPE | STRENGTH IN 1000 PSI | | | MODULUS OF ELASTICITY, TENSION $\times 10^6$ | ELON-GATION, % IN 2 IN. |
	TENSION STRENGTH	YIELD STRENGTH	SHEAR STRENGTH		
260:					
¼ hard	54	40	36	16	43
Hard	76	63	44	16	8
360:					
Hard	68	52	38	14	18
377:					
Annealed	52	20		15	45
521:					
Hard	93	72		16	10
752:					
Hard	85	74		18	3
11A	40–50	20–30		19	15–25
9B:					
Non-heat-treated	65	30		16.5	15
HT	85	48		17	9

and resist both wear and fatigue. Type 521 is representative and is widely used. Its properties are given in Table 3-12.

A group of copper alloys containing considerable nickel is called nickel silver. They are non-heat-treatable. The cold-working qualities are excellent, but hot working is difficult. Type 752 is typical of the wrought form, and is most often used. The uses include a wide variety of hardware and small intricate parts such as those used in cameras. Alloy 11A is representative of the cast form and is used for such purposes as hardware, building trim, and marine fittings. The properties of both are given in Table 3-12.

The bronzes that can be heat-treated are aluminum bronze and beryllium copper. Aluminum bronzes are used in both the cast and wrought forms. The characteristics and properties can be varied over a wide range by changes in the amount of alloying elements, and for some types by heat-treatment. These alloys have good corrosion resistance and may be worked hot or cold. They are resistant to both shock and fatigue even as castings. Alloy 9B is a general-purpose aluminum-bronze casting alloy with many applications, including machine parts and chemical and marine equipment. Properties for both the non-heat-treated and the heat-treated conditions are given in Table 3-12. Beryllium copper is also a very versatile alloy that has a wide variety of uses, from bolts to surgical instruments. It has excellent formability in the soft condition, and high fatigue strength and creep resistance in the hardened condition. It has good corrosion resistance and electrical conductivity. Annealed, the tensile strength is 60,000 psi, and the elongation is 35 percent (percentage in 2 in.). It can be heat-treated to a tensile strength of 200,000 psi; the elongation is then 2 percent.

3-42 MAGNESIUM ALLOYS

Magnesium alloys are used in many applications where weight is an important factor. These alloys are the lightest of the commercially used metals, their weight being only one-fourth that of steel. However, their cost is roughly 10 times that of steel; thus, they should be specified only in applications where weight reduction is of the utmost importance. These alloys have good resilience, damping qualities, and fatigue strength, though they are very sensitive to stress concentrations. There is a considerable reduction in strength at a temperature of only 200°F.

Magnesium alloys are noted for their machinability. Only about one-half as much power is required as for machining aluminum or brass. Tool life is several times longer than for other metals, and the accuracy and surface finish are very good. Fire is a hazard in machining, but with care it is not a disadvantage of any consequence. Any of the casting processes can be used, and forging causes no difficulty, but cold forming does;

TABLE 3-13 Properties of Miscellaneous Metals

| MATERIAL | STRENGTH IN 1000 PSI | | | ELONGATION, % IN 2 IN. |
	TENSION STRENGTH	YIELD STRENGTH	COMPRESSION STRENGTH	
Magnesium:				
Alloy AZ31B	40	25	14	16
Alloy AZ91C-T6	40	19	19	5
Zinc alloy, ASTM				
AG40A (SAE 903)	41		60	10
Monel:				
Cold-drawn bar	100	85		20
Inconel:				
Cold-drawn bar	120	100		22

however, deep drawing can be accomplished with heated dies. Arc welding is the type of welding most often used. If due consideration is given to surface treatment and assembly, corrosion is not a great disadvantage.

The properties for two frequently used magnesium alloys are given in Table 3-13. Alloy AZ31B is used for sheet, plate, extrusions, and forgings. Alloy AZ91C-T6 is used for sand castings and permanent-mold castings.

3-43 ZINC ALLOYS

One of the most extensive uses of zinc alloys is for die castings. They are easy to cast and can be used for intricate castings and for very thin sections. The tolerances can be close and the finish is good; they can be easily machined, and they retain sufficient ductility so that they can be bent, if necessary, to produce the part more easily. They have greater strength than aluminum at room temperature and are about 30 times as resistant to atmospheric corrosion as is plain carbon steel. They cost roughly 3 times as much as plain carbon steel.

Zinc alloy die castings are difficult to weld or to solder. They have little strength at temperatures higher than 200°F and should not be used for low temperatures, because of brittleness. They are used in a great many products where the quantity produced is large. The properties of a representative grade are given in Table 3-13.

3-44 NICKEL ALLOYS

Commercially pure nickel is seldom used as a metal from which parts are manufactured, but there are a number of nickel alloys whose properties and characteristics permit their use in some very demanding applications. They possess high strength and toughness, and excellent resistance to many corrosive agents, both at normal and elevated temperatures. For some, the operating temperature approaches 2400°F. Most maintain their strength and toughness at low temperatures, some as low as −400°F. Fatigue strength and resistance to creep are also outstanding.

Nickel alloys are rather difficult to machine. They are readily formed in the annealed condition, and they can be welded with usual methods. The casting qualities are similar to those of steel.

The more common alloys are known as monel and inconel. The average properties of a representative grade of each of these alloys are given in Table 3-13.

3-45 MISCELLANEOUS METALS

Titanium combines, to an extent, the desirable characteristics and properties of aluminum and steel. Most titanium alloys are suitable for service at operating temperatures approaching 1000°F and have excellent corrosion resistance. Other metals have superior heat resistance and equal corrosion resistance, but they do not have the desirable strength-weight ratio of titanium. Titanium is also used at operating temperatures less than −400°F. Tending to offset these advantages are the difficulties of fabrication and cost, which may be over 10 times as great as for alternative metals.

Cobalt alloys are noted for their great hardness, corrosion resistance, wear resistance, resistance to galling, and magnetic properties. For some alloys, these characteristics are to be had at temperatures over 2000°F. Manufacturing involves no unusual difficulties. Because of the cost of the alloys, they are often used as a facing; that is, a layer is deposited by welding to the areas of a part that require the great resistance. One of the most important uses is for cutting tools.

Lead is used for many purposes. Its important characteristics are: resistance to corrosion, the ease with which it is worked, high density, low melting point, and low cost. However, lead has little strength, and to avoid creep at room temperature the stress should not exceed 200 to 300 psi. Some of the less obvious uses are: bearing plates for bridges, radiation shielding, and sound barriers.

Other metals such as molybdenum, tungsten, and beryllium have desirable characteristics or properties, but their use is restricted by disadvantages, such as: brittleness, tendency to oxidize, or manufacturing difficulties. Metals such as gold, silver, and platinum, known as precious metals, are used in a number of unique applications. Their very high cost is only one of the disadvantages that limit their use.

3-46 PLASTICS

Plastics are useful in some instances in mechanical design, because of the low cost of the finished part, or because of the unique characteristics or properties of plastics. In general, plastics are lightweight, good electrical insulators, and good heat insulators. Some are resistant to corrosion; others are transparent or translucent, and some may be produced in any color.

Plastics fall into two basic classifications: thermoplastic and thermosetting. The thermoplastics are those that may be repeatedly softened by heating to a relatively low temperature. The thermosets are those that undergo a chemical change when the part is manufactured and cannot be softened. There are a number of basically different kinds, and each has several, in some cases many, different types or grades, the properties or characteristics of which are considerably different.

Plastics are available in many forms: film, foam, extruded, cast, and molded. Foams may be flexible, rigid, or foamed in place. Extrusions may be rigid or flexible. Cast or molded plastics may have one of a large variety of filler materials added to greatly increase some desirable characteristic or property. The initial process of manufacture may produce a finished part, a semifinished part, or merely stock, which is processed by common methods such as stamping or machining.

The thermosetting plastics, in general, are stronger, harder, and have greater resistance to heat and chemicals. The thermoplastics are less brittle and are more readily molded.

Plastics do not have the strength of the structural metals, but when compared on a strength-weight basis, they are more favorable than consideration of the strength alone would indicate. Plastics have limitations that must be given consideration when deciding whether to use a plastic and in selecting a particular type of plastic. Some of these limitations are: creep, moisture absorption, thermal expansion, flammability, embrittlement at low temperature, low ductility, poor resistance to sunlight, cold flow, and poor thermal conductivity. Not all these limitations are important in every application, nor do they apply to every plastic. Plastics generally have low material and processing costs.

Phenolics have excellent electrical properties and strength. They are resistant to heat, moisture, and many chemicals. They are lightweight and can be produced with a smooth finish. They can be molded into intricate shapes, they retain their dimensional stability for a long time, and they are self-extinguishing. Phenolics are the oldest and most-used thermosetting plastic. Bakelite is the original phenolic plastic. There are about 20 different types, the properties of which do not vary greatly. Phenolics are limited to only dark colors, and they should not be machined except

TABLE 3-14 Properties of Plastics

PROPERTY	PHENOLIC	NYLON
Density, lb/cu in.	0.05	0.041
Tensile strength, psi	7000	11,000
Modulus of elasticity in tension, psi	1.5×10^6	420,000
Water absorption, % increase in weight in 24 hrs at 73°F	0.6	1.5
Maximum continuous operating temperature, °F	300	250
Compressive strength, psi	30,000	15,000
Coefficient of thermal expansion, 10^{-5} in./(in.)(°F)	2	5

when it is impossible to mold the desired configuration. Applications include: camera cases, electrical parts, instrument cases, bottle caps, electrical appliance handles, ignition system parts, radio cabinets, pulleys, and control knobs. The properties given in Table 3-14 are for the general-purpose type, the cost of which is about 1.5 cents/cu in.

The representative thermoplastic chosen is nylon. There are several types. They are tough, abrasion resistant, and are resistant to some chemicals. They are readily molded and will flow around complicated inserts. They also have good resistance to creep and are self-extinguishing. Cost is about 4 cents/cu in. Nylons can be produced in colors, or they can be dyed. Parts can also be produced by machining stock.

The characteristic of nylon to absorb moisture must be considered when close tolerances and an environment where moisture can be absorbed are involved, because of both the change in size and reduction of strength. The properties for type 66, the most common, are given in Table 3-14. Applications include many mechanical components, such as gears, cams, and housings for electrical appliances. Nylon is especially well-suited for use as a bearing material, either with or without lubrication. A few typical nylon parts are shown in Fig. 3-5.

Fluorocarbons are not as strong as nylon, they absorb little or no water, and they are inert to most chemicals. The operating temperature for some types is twice that of nylon. Another outstanding characteristic is that virtually nothing can adhere to the surface. Teflon is a fluorocarbon. The cost is much greater than for nylon — for one type over 10 times as much. Applications are mechanical parts that require resistance to chemicals or to heat.

Polyethylene is the most used of the thermoplastics. There are a great many types. The strength and heat resistance is greatly inferior to nylon.

Fig. 3-5 Typical nylon parts. (*E. I. du Pont de Nemours & Co.*)

The moisture absorption is negligible, and most types will burn. The cost is about 1 cent/cu in. Typical applications are containers, toys, pipe, and mechanical parts.

3-47 MISCELLANEOUS MATERIALS

Elastomer is the term used for a large number of materials that are generally thought of as rubber. The properties or characteristics that are often responsible for the choice of an elastomer are as follows:

1.   Resilience — The tendency to snap back.
2.   Hardness — Should not be interpreted as stiffness.
3.   Abrasion resistance — May not be required for a particular application.
4.   Electrical properties — Most elastomers can be made acceptable insulators.
5.   Damping and impact resistance — Excessive temperatures can be built up by rapid cycling.

6.   Impermeable to liquids or gases — Most elastomers can be made so as to provide an adequate barrier.

Typical applications that require some of the above characteristics or properties are: hose, bushings, electrical insulation, belts, seals, and diaphragms.   An elastomer is also an essential component in some shock mounts, vibration dampers, and flexible couplings.

Factors that affect the choice of a particular elastomer are:

1.   Heat — The maximum permissible operating temperature is usually around 200°F, though for some it is 600°F.

2.   Cold — The minimum permissible operating temperature is usually about −20°F, though for some it may be as low as −150°F.

3.   Sunlight, weather, fuel, and oil — These elements are generally detrimental to elastomers, but some can be made that have adequate resistance.

4.   Adhesion to metal — This can usually be accomplished without difficulty.

5.   Cost — Generally, there is little difference between the various types, though some may cost 25 times as much as the average.

The manufacturer can tailor an elastomer to specific requirements, within limits.   However, it must be realized that the characteristics and properties are interdependent.   For example, an increase in hardness to improve wear resistance will be accompanied by a decrease in resilience. The choice invariably involves a compromise that should be based on careful consideration of all the factors or with the help of a supplier.

A number of other nonmetallic materials used in a relatively few mechanical applications are: glass, ceramics, mica, felt (for shields, retainers, gaskets, sound deadening), wood, industrial paper, laminates (for electrical parts that must be nonconductors and possess stiffness), and carbon-graphite.

Because of its importance in mechanical design, carbon-graphite will be discussed briefly.   There are a number of different types.   They can be all carbon, or all graphite, or a combination of the two.   Also, other material may be added to produce some particular characteristic or property.   The unique combination of characteristics of carbon-graphite are: low coefficient of friction without lubrication, low coefficient of thermal expansion, and high thermal conductivity.   In addition, it is inert to most chemicals, the operating temperature may be as high as 1200°F, and it can be ground to a very close tolerance.   The strength varies greatly with the different compositions and is much greater in compression than in tension.   Common applications are: bearings, seals, and pump parts.

QUESTIONS

3-1 On what basis should a metal be selected for a part that must adapt itself to a slightly different configuration when subjected to a high stress?

3-2 What property of metals is used to evaluate their suitability for use in a part that requires great stiffness?

3-3 What is the difference between resilience and toughness?

3-4 What is the difference between fatigue strength and endurance limit?

3-5 Elongation is one important property in determining the formability of a metal. What is another important property?

3-6 What are two important factors that determine the castability ratings of metal?

3-7 What are some of the desirable characteristics of gray cast iron?

3-8 What is the outstanding characteristic of white cast iron?

3-9 What are some of the important advantages of nodular iron castings that justify its use even though it costs more?

3-10 What unusual characteristic does gray cast iron have when subjected to high temperature?

3-11 How can the strength of low-carbon steels be increased?

3-12 What are the two methods of cold-finishing steel?

3-13 What is done to make a steel free-cutting?

3-14 What is one of the major reasons for using a medium-carbon alloy steel instead of a plain carbon steel of the same carbon content?

3-15 Discuss the basis upon which stainless steel is classified into three groups.

3-16 Why do tool steels cost much more than other steels?

3-17 Are steel castings as easy to produce as iron castings?

3-18 What are two common ways of dealing with the inferior corrosion resistance of the high-strength aluminum alloys?

3-19 In the designation 1100-H14 what does the 4 mean?

3-20 Explain why a given non-heat-treatable alloy may be obtained in a harder temper if it is 0.25 in. thick than if it is 2.0 in. thick?

3-21 Why does an alloy produce greater strength as a permanent-mold casting than as a sand casting?

3-22 Why are copper alloys seldom used for die casting?

3-23 Can both brass and bronze be heat-treated to improve their strength?

3-24 Explain how the properties and characteristics of aluminum bronze can be varied over a wide range.

3-25 What are several advantages of magnesium?

3-26 What are several advantages of nickel alloys?

3-27 Discuss the basis upon which plastics are classified into two groups.

3-28 What are some of the important limitations of plastics?

3-29 What are the properties or characteristics that are often responsible for the choice of an elastomer?

3-30 What are the unique characteristics of carbon-graphite?

SELECTED REFERENCE

"Metals Properties," ASME, New York, 1954.

4 MANUFACTURING PROCESSES IN DESIGN

4-1 INTRODUCTION

There is a tendency for a designer to take the attitude, "My job is to design it so it will work; how it is to be made is someone else's worry." But unless careful consideration is given to the manufacturing processes the other work is a fruitless exercise, for a part or machine that is impractical to manufacture is as much a poor design as one that will not operate correctly.

The design dictates the manufacturing processes to be used to a greater extent than is generally recognized. By the time the conditions such as stress, vibration, corrosion, and abrasion have been provided for, and the configuration including tolerance and surface finish has been worked out, and the material and type of heat-treatment have been selected, nearly all the manufacturing processes have, unintentionally, been determined. To a great extent, efficiency in production is something that is achieved in the design stage, and it should be the aim of the designer to design so that, consistent with other requirements, the product will have the necessary attributes when produced in the most efficient manner for the quantity that is required. A designer can give the manufacturing processes due consideration only if he is familiar with the capabilities and limitations of the various processes.

It is the purpose of this chapter to acquaint the beginning designer with the characteristics of manufacturing processes from the point of view of the designer. He can then determine the most likely processes that could be used in performing an operation on a part and

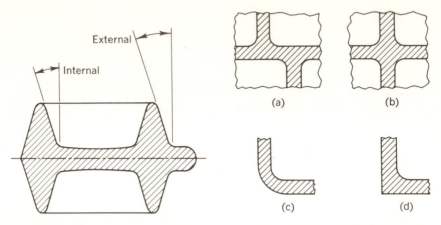

Fig. 4-1 External and internal draft.   Fig. 4-2 Treatment of intersections.

then design within the limits of these processes; he can design for the minimum number of different processes; and he can so design that standard machines and cutting tools can be used.

This background will also qualify him to give a considered and reasonable answer to the two basic questions that pertain to the relation of manufacturing processes to design: Will the design permit production in the most efficient manner? And will the most economical processes produce the required quality — strength, hardness, tolerance, surface finish, etc.?

It should be clearly understood that it is not recommended here that the designer specify the manufacturing processes, or that he be given or accept the responsibility for determining the method of manufacture, but only that he acquire the knowledge essential to adequately perform his function as a designer.

4-2 CASTING

Any metal that can be melted and poured can be cast. The casting processes permit a very wide latitude in design. However, there are concepts and details with which a designer must be familiar before he can produce a successful design. There are a great many variables in casting, including the skill of those making the castings and the amount of care they are required to take in doing their work.

Wherever possible, the parting plane should be flat and located so that the largest dimension of the part is in this plane. It is desirable that all mold cavities be as shallow as possible, and consideration should also be given to the necessity of placing cores in the mold. It would be well if the parting

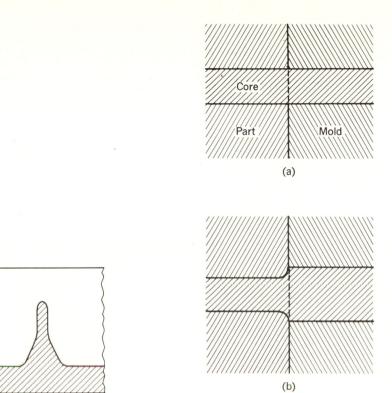

Fig. 4-3 Wall transition at intersection.

Fig. 4-4 A radius at a cored hole is undesirable.

plane were kept away from surfaces that would be likely to be used for locating or clamping when the part is machined.

All surfaces perpendicular to the parting plane require draft to permit easy withdrawal of the pattern. External draft is that used at surfaces which on cooling will shrink away from the mold, and internal draft is that used on surfaces which on cooling tend to tighten on the mold. This is illustrated in Fig. 4-1. Cores must have draft to permit withdrawal from the core box.

All parts of a casting should be as close to the same thickness as possible. Any heavy sections should be on the outside and at the parting plane, so that they can be fed directly. Intersections should be so designed that concentration of metal is at a minimum. Figure 4-2a and c illustrate good design and Fig. 4-2b and d illustrate poor design. Whenever a thin wall intersects a thick wall, the transition should be as shown in Fig. 4-3.

An effort should be made to design a casting and to locate the parting plane in a manner such that cores are not required. However, they are often used to advantage. A core should be adequately supported at both ends. A

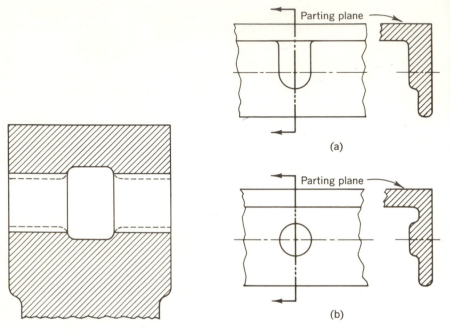

Fig. 4-5 Arrangement of a core to reduce machining.

Fig. 4-6 Treatment of bosses.

corner radius should not be required at a cored hole. Figure 4-4a shows the correct method. When a cored hole is also to be machined, the machining can be reduced if the core is made as shown in Fig. 4-5.

A fillet radius should be required, to avoid a stress concentration and to prevent the formation of a crack that often occurs during cooling when a sharp corner is present. However, this radius should not be too large, for this would create an undesirable concentration of metal. A corner radius should be used for all corners except those at the parting plane.

Bosses that are not parallel to the parting plane should have the form shown in Fig. 4-6a, rather than the circular form shown in Fig. 4-6b, which requires a core. If the boss is to be spotfaced, the radius should be smaller than the radius of the spotface. The height or thickness of the boss should be great enough that when it is machined there will still be some boss remaining as shown in Fig. 4-7a, rather than as shown in Fig. 4-7b. Mounting pads should be provided, as shown in Fig. 4-8a, rather than as shown in Fig. 4-8b.

When designing a part to be made by casting, careful consideration should be given to any machining that must be performed. Machined surfaces should never be arranged to match or blend into cast surfaces.

Metal must be allowed on surfaces that are to be machined; it is referred to as machining allowance. Though it is not universally agreed that the machining allowance should be specified on the drawing, it is necessary for the designer to know the approximate amount in order to design the part. In determining the machining allowance, consideration must be given to the following: parts of the mold may not be perfectly aligned, cores may shift from the desired location, draft is necessary, there must be a minimum of about 0.03 in. of metal to remove, and there must be a tolerance on both the casting dimensions and the machining dimensions.

Castings require relatively large tolerances. Some of the factors that are responsible are: different amounts of shrinkage caused by variations in pouring temperature, incomplete closing of the molds, rapping the pattern to make it easier to remove it from the mold, and of course, the tolerance that must be allowed on the patterns or molds. In the design of parts where castings are involved, the large tolerances must be allowed for in providing clearances.

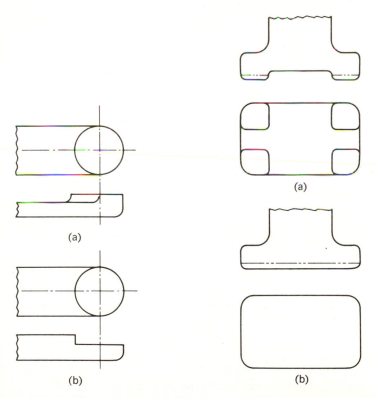

Fig. 4-7 Provision for machining a boss.

Fig. 4-8 Treatment for mounting surfaces.

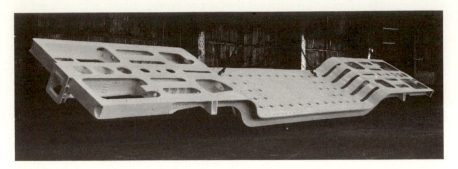

Fig. 4-9   A steel casting. (*Steel Founders' Society of America.*)

Sand casting is the most common process for producing castings. Because the molds and cores are destroyed, limitations on the design to allow for their removal do not exist. The process seldom limits the size of a proposed part. Steel castings of over 200 tons are not unreasonable. The tooling costs are low; thus, the production of only a few parts is feasible. The railroad car frame shown in Fig. 4-9 is a single steel casting.

The minimum thickness that may be used depends on the metal to be cast and the size of the casting. However, sand casting does not permit as small a thickness as other casting processes. An appropriate minimum thickness for comparison with other processes is 0.20.

The desired external draft is 2°, though ½° may sometimes be used. Internal draft is usually twice as large as external draft. The comment on minimum thickness also applies to fillet and corner radii. An appropriate fillet radius for comparison is equal to the minimum thickness, and the corner radius is one-half as large. An allowance for mold mismatch need not exceed 0.03 for a moderate-size part, and the same may be said for core shift. Cast-in inserts are not practical in sand casting.

Reasonable tolerances for aluminum, magnesium, and malleable iron are: ±0.03 for dimensions to 6 in., and ±0.003 per in. additional for each inch over 6; for steel: ±0.06 to 12 in., and ±0.005 additional for each inch over 12. For cast iron the tolerance may be slightly less than for steel, and for copper alloys it should be slightly greater than for steel. The angular tolerance is ±½°. The surface roughness is generally from 500 to 1000 μin., though with special treatment a considerably smoother surface can sometimes be produced.

Die casting requires a large machine and metal molds. Thus, from 1000 to 5000 parts, depending on the complexity, are required to justify using the process. The metal molds also impose limitations on the design. The machines most often available limit the size of a casting to about 30 lb for aluminum. The life of a die using aluminum is about 300,000 parts per cavity, using copper alloys about 15,000, and using zinc the life is almost unlimited. The die casting of ferrous metals is now in the development stage.

All walls should be as uniform as possible. The preferred thickness is 0.10, the minimum 0.06, and the maximum about 0.20. Thicker sections will have considerable porosity and voids. For smaller parts, the thickness can be less than as specified above. And it should be noted that zinc permits thinner walls than does aluminum.

The preferred minimum draft is 2° for external and 3° for internal. Corner radii should be 0.06 where possible, and never less than 0.03. The fillet radius should be approximately one-third the sum of the two adjacent wall thicknesses.

Inserts may be used. Consideration must be given to providing for adequate locking, and the metal should contract toward the insert. Threads can be cast; however, external threads smaller than ¾ in. or internal threads smaller than ⅜ in. are generally machined.

Reasonable tolerances on dimensions in the same half of the die are ±0.008 for the first inch and ±0.0015 for each additional inch. For dimensions across the parting plane, an additional ±0.010 is required. An allowance of at least 0.010 should be made for mismatch. A surface roughness of 63 μin. can be produced in production.

Figure 4-10 shows a large aluminum die casting being removed from the die casting machine. Both halves of the mold are clearly visible.

Permanent-mold casting is the same as sand casting except that a metal mold is used. Because the mold is not destroyed, it is less versatile than sand casting. The tooling costs are two to three times as great as for sand casting; thus, over 500 castings are generally required to justify the process. The maximum size is about 50 lb for aluminum.

The minimum thickness is 0.12 for small areas. The desired corner and fillet radii are 0.06 and 0.18. The preferred minimum draft is 2° external and 4° internal. The minimum tolerance between points in the same half of the mold is ±0.01 to 6 in., plus ±0.002 for each additional inch of length. Across the parting line the tolerance should be increased 50 percent. The machining allowance for moderate-size parts is 0.06. The surface finish is usually about 200 μin.

Plaster-mold casting permits controlling the thermal characteristics of the mold, which in some instances is highly desirable. The mold and cores are destroyed; thus, the process is very versatile. The tooling costs are less than for permanent-mold casting, and it is not a quantity-production process. It is best suited to the production of relatively small, nonferrous castings, though plaster-mold castings weighing over 1 ton have been made.

The minimum thickness is 0.04 for small areas, but 0.06 is preferred. The preferred fillet and corner radii are 0.06 and 0.03, though one-half this is possible. The desired draft is 2° external and 3° internal, though ½° can be

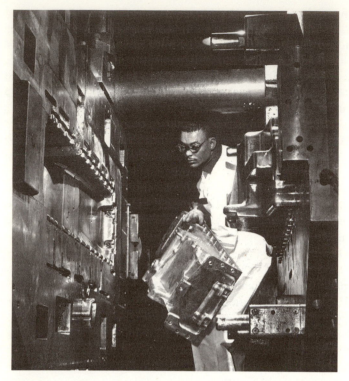

Fig. 4-10 An aluminum die casting. (*Aluminum Company of America.*)

achieved. The tolerance between points in the same half of the mold is 0.005 per inch of length; across the parting line it is twice as great. A surface roughness better than 100 μin. is usually produced, and machining can often be omitted.

Shell-mold casting employs a thin shell-like mold of sand and resin. The cost is higher than the usual sand-casting process, and thus should not be used unless the reduction in machining or the elimination of cores can offset this higher cost. Often, as few as 100 castings can justify using the process. Castings produced by this process seldom exceed 50 lb for aluminum or 70 lb for steel.

The minimum thickness varies with the metal. The preferred minimum for aluminum is 0.12, for magnesium and bronze 0.16, and for steel 0.19. The desired draft is 1° external and 2° internal, though one-half these values can be achieved. It is especially important to have a flat parting plane. The desired corner and fillet radii are 0.06 and 0.12. The machining allowance is 0.03 on small parts to 0.06 on moderate-size parts. The minimum tolerance is ± 0.006 per in. for small dimensions in the same side of the mold, and ± 0.04 for dimensions over 8 in. An additional ± 0.01 is required for dimensions across the parting plane. The surface roughness is generally over 100 μin.

Investment casting is well suited to producing intricate parts. When wax patterns are used, the tooling cost is not especially high. Wax patterns are most often used. The cost per casting is the highest of any of the casting processes, because of the great amount of handling that is required. This process is seldom used for castings weighing over 15 lb in aluminum or 40 lb in steel. It is indicated for small parts that would require considerable machining if made by other processes or that would be difficult to machine.

The desired minimum thickness is 0.04. Thinner sections can be produced. The preferred corner and fillet radii are 0.03 and 0.06. The preferred draft is 1° external and 2° internal, though surfaces with no draft can be produced. The minimum tolerance between points in the same side of the mold is ± 0.005 per in.; for dimensions across the parting line ± 0.005 must be added. A surface roughness better than 100 μin. can be produced.

Occasionally, centrifugal force is used to improve the quality of castings. The molds may be sand, plaster, or metal. Such castings are often referred to as centrifugal castings. A true centrifugal casting is one which is symmetrical about the axis of rotation and which has a rather large hole in the center, such as a pipe or a wheel. Except for consideration of the centrifugal force, the material used for the mold is the controlling factor. A point which the designer should keep in mind is that air pockets and impurities will be near the axis of rotation, and if they would be objectionable in the part, provision must be made to remove them by machining.

4-3 FORGING

Forging is the process in which metal in a plastic, rather than a molten, state is forced to flow into the desired configuration. This flowing produces an improvement in the properties of the metal, and the accurate shaping reduces the machining that is required to make a part. However, the expensive dies limit the process to quantity production. Though die forgings weighing several tons are made, three-fourths of the forgings produced weigh less than 2 lb.

When designing a part to be produced by forging, advantage should be taken of the grain; that is, tensile and impact loads should be parallel to the grain, and shear loads should run across the grain. Also, machining that cuts deeply into the forging should be avoided.

Figure 4-11 illustrates the complexity that can be achieved by forging. This aluminum forging becomes part of the hydraulic system of the F-111

Fig. 4-11 A complex aluminum forging. (*Wyman-Gordon.*)

supersonic fighter aircraft. Over 1200 dimensions were involved in specifying the configuration.

A cavity in only one die is desirable. When this is not feasible, the

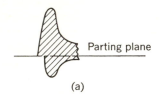

(a)

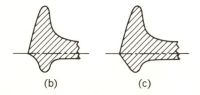

(b)   (c)   **Fig. 4-12 Methods of matching draft.**

parting plane should divide the part into approximately equal portions. Any webs should have the parting plane pass through their center, and the parting plane may be stepped to accomplish this. Deep, narrow cavities should be avoided.

In providing for draft, it would be well to keep in mind that it is produced by taper on the cutter used to make the die, and that the part is dimensioned as though there were no draft. The angle should be constant. For steel, the standard angle is 7° for external draft and 10° for internal. For aluminum, 5° is often used, and in some instances as little as 1° is feasible, but the cost will be considerably greater. When the draft in one die does not quite match the draft in the other die, as shown in Fig. 4-12a, a rather large radius should be used to cause them to match, as shown in Fig. 4-12b, or the draft angle should be increased, as shown in Fig. 4-12c.

Fillet radii should be as large as possible. A large radius facilitates the metal flow, produces a more uniform deformation, reduces the force required to make the forging, and reduces the cost. The corner radius should be large, because in the die it is a fillet radius, and if it is too small a stress concentration is produced which will cause premature die failure. This radius should be constant, to permit using the same cutter for the entire die. The appropriate size of the corner and fillet radii depends not only on the size of the forging but on the depth of the cavity in the die. For a moderate-size steel part and a depth of 0.5, the minimum corner radius is 0.06; for a depth of 1.0, it is 0.09; for 2.0, it is 0.13. The preferred fillet radius is three times as great as the corner radius. For an aluminum forging, these radii should be slightly greater.

Thin webs should be avoided, because they cool quickly and require excessive hammering to produce them. The minimum thickness is largely dependent on the area of the web. For a steel forging with a web area of less than 20 sq in., the minimum thickness is 0.13, and for a 100-sq-in. web, it is 0.23. For an aluminum forging, the thickness should be slightly greater.

The depth that a boss extends into the die should not exceed two-thirds of its diameter or width. There should be a minimum of 0.03 between the finished surface of a boss and the adjacent rough forging. The thickness between bosses can be coined to ±0.005, even though the thickness is across the parting plane. The usual surface roughness is 250 to 500 μin. Smoother surfaces can be produced, especially with aluminum.

There are several tolerances related to forgings. Thickness tolerance applies to dimensions across the parting plane and is largely dependent on how close the dies come to closing perfectly. Length tolerance, also referred to as length and width tolerance, is the required tolerance between points in the same die and is determined by shrinkage and the allowance for the wear that will take place in the die. There is also a mismatch tolerance that has the same meaning as it does for castings. The draft angle is usually understood to have a considerable tolerance. For parts that have a tendency to warp, a straightness tolerance is necessary. Not all these tolerances, of course, apply to each dimension of the forging. Mismatch would not apply to dimensions perpendicular to the parting plane or to dimensions in the same die. It should be remembered that, though aluminum has a larger coefficient of thermal expansion, steel is forged at a higher temperature and has a greater plastic range.

The term machining allowance is sometimes used to indicate the stock that must be removed in machining to ensure that the surface will properly clean up. More often, it is used to indicate not only the stock to be allowed for cleanup but the metal required to provide for all the tolerances that apply. For a forging of moderate size and normal configuration, providing for machining in this manner is practical. For steel forgings less than 1 ft long and weighing less than 10 lb, a reasonable machining allowance is 0.12. For aluminum forgings less than 1 ft long and weighing less than 3.5 lb, the allowance is 0.09. For small forgings of either metal, the allowance can be less. For larger forgings, the machining allowance is calculated. Not less than 0.03 of material should be provided for cleanup in addition to that which must be allowed for the various tolerances.

Hot upsetting is a process used to form a head on a heated rod by shortening the rod. The shape of the head can be provided by a cavity in either the ram or the split dies used to grip the rod. Because of the dies required, this is a quantity-production process, though for parts that do not require cavities, the number of parts need not be large. Some machines can handle stock 9 in. in diameter. Steel is the most satisfactory metal.

The maximum length that can be upset into a head in one stroke is determined by the buckling of the unsupported rod. This length is equal to about 3 times the rod diameter. If the diameter of the upset head is not greater than 1½ times the diameter of the rod, a length much greater than three diameters can be upset in one stroke. When a larger quantity of metal is required in the head, several strokes with different dies can be employed. The process is not limited to gathering metal only at the end of the rod. The gathering may take place anywhere along the rod, but the dies required are more complex and thus more expensive. The length limit of three diameters should not be exceeded.

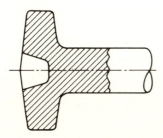

Fig. 4-13 Draft on a hot upset part.

Wherever possible, draft should be permitted to facilitate the flow of metal and the removal of the dies and ram from the part. This is illustrated in Fig. 4-13, where the draft angles have been exaggerated. The desired minimum external draft is 3°, and the internal is 5°. The ram may be used to pierce the head to produce a hollow part. Corner and fillet radii equivalent to those of a die forging are appropriate.

The wall thickness of tubing may also be increased by this process. The increase in wall thickness which produces a larger outside diameter is limited to about one-fourth the tube wall thickness. When the increase in wall thickness reduces the inside diameter, the increase in wall can be much greater.

Tolerances equal to a die forging of similar size can be held. A machining allowance of 0.06 to 0.10, depending on the size, should be provided.

Cold heading is very similar to hot upsetting, except that the stock is not heated, and therefore, it is more limited with respect to size and complexity of parts. However, the fact that the stock is not heated permits much closer tolerances and a smoother surface finish; it also costs less than hot upsetting. Those metals whose strength is increased by cold working are made stronger by cold heading. The minimum number of parts to justify the process is over 5000 and may be 25,000, depending on the complexity of the part. Aluminum, copper alloys, and steel are all suitable. Steel with less than 0.15 percent carbon content is the most commonly used.

The maximum diameter that is practical is about 1 in.; the length should not exceed 10 in. for 1 in. diameter and should be proportionately less for smaller diameters. The length of stock to be used in forming the head should not exceed 6 times the diameter. And if it is less than twice the diameter, a simpler machine can be used. The maximum diameter for stainless steel is about 0.50 in.

The preferred minimum tolerance for diameters and between points in the same die cavity is ±0.002, though twice this will greatly increase the die life and thus reduce the cost. The tolerance on the overall length should be ±0.03 wherever possible.

Cold drawing is the process in which an unheated rod is pulled through a die to produce the desired cross section. Shapes somewhat similar to those produced by hot extrusion may be made, though they must be much simpler. Steel is often used. The maximum size is about 7 sq in. of cross-sectional area and a width of 4 in. This process requires a rather large quantity to make it economical.

Fillet and corner radii should be used, though they can be relatively small. The surface is smooth and should be used as drawn. Cold working improves the wear resistance of the surface. A tolerance of ±0.005 is reasonable, though small sections can be held closer.

Swaging is used to reduce the diameter of a solid rod or a tube. Though the process can be performed on either hot or cold stock, heating should be

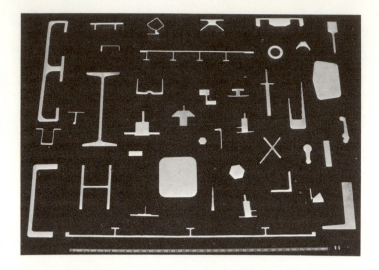

Fig. 4-14 Aluminum extrusions. (*Harvey Aluminum.*)

avoided where possible. Equipment is generally available to handle tubes to 6.5-in. diameter and rods of somewhat smaller diameter. The included angle of the taper between the two diameters should not exceed 30° for tubing and 8° for solid rod. The maximum length of taper produced in a single operation is about 1.5 ft. The tolerances on external dimensions and surface roughness are good; however, when an accurate inside configuration is required, a mandrel must be used.

Hot extrusion is the process by which metal heated to the plastic state is forced through a die to produce the desired shape of cross section. Cross sections that are impossible to produce by rolling sheet metal are easily achieved. Aluminum is particularly well suited to this process. Figure 4-14 shows the cross sections of a number of aluminum extrusions.

The first thing to do when considering an extrusion is to determine whether a section for which a die is available can be used. The cost of a die for a small simple section is low enough that several hundred pounds of extrusion is all that is needed to justify producing a new section. Machines are usually available to make a section that can be placed in a 16-in.-diameter circle. The maximum length varies from 10 to 40 ft, depending on the size of the section.

The minimum thickness depends on the metal and the size of the section. Larger sections require a greater thickness. For small sections the minimum for aluminum is about 0.04, and for copper alloys it is about 0.12. The preferred minimum corner radius is 0.03. Sections that require the die to have a long, slender tongue should be avoided.

When aluminum or magnesium is used, a completely enclosed section can be produced. The center portion of the die is supported by thin webs, which cause the metal to part and then weld itself together after passing the webs but before leaving the die.

Tolerances that can be held vary considerably with the section and size. However, the following give an idea of what can be expected: for dimensions over 2 in., ± 0.020 plus ± 0.005 per in. for each additional inch; dimensions less than 2 in. slightly closer; angles $\pm 1\frac{1}{2}°$; straightness ± 0.012 in. per ft; twist $\frac{1}{2}°$ per ft, but not more than $5°$. With care, these tolerances can be considerably reduced, but the cost will, of course, be increased.

Impact extrusions are cup-shaped parts made by placing the unheated stock in a die and striking it with such intensity that it flows between the die and the punch. There are two methods: forward extrusion and backward extrusion. The forward method can produce longer parts, but it is used less extensively. Only the backward method will be discussed further. When the stock is struck, it is forced back up the punch. The length of the part is thus limited by the strength of the punch as a column. Aluminum is the most extensively used metal, though even steel can be used. The maximum size for aluminum is about 5 in. in diameter and 15 in. long. The process is suitable for moderate and very large quantities.

The length should not exceed 10 times the inside diameter. The minimum wall thickness varies from about 0.004 for small-diameter parts to about 0.08 for larger parts. The thickness of the bottom should be $1\frac{1}{2}$ times the maximum wall thickness. A round or slightly beveled bottom is preferred to one that is flat. The process is not limited to circular parts, but for maximum accuracy they must be symmetrical, and the area of the punch should not exceed 20 sq in. The preferred tolerances are ± 0.04 for diameter, length, and bottom thickness, and ± 0.02 for wall thickness. All except the length can be held considerably closer, if necessary, especially with soft metals.

4-4 TURNING AND BORING

Turning is done on a lathe or a machine that is a refinement of a lathe. The parts are often made from stock of circular cross section, but it must be remembered that castings and forgings can be held in a chuck or a special fixture to permit their being machined.

An engine lathe is the basic turning machine. It is very versatile with respect to both the operations it can perform and the size of the parts that can be produced. It can machine parts of almost unlimited size. It is well suited to the production of a few parts, but other turning machines are more appropriate for quantity production. Parts that have a length in excess of 15 times their diameter require special precautions because of the deflection caused by the force of cutting, and the cost is, therefore, greater. The concentricity of the various machined diameters can be held closer when they are all machined in a single setup. All fillet radii that can be made by the same cutting tool should have the same dimension. Corners should have a chamfer rather than a radius. If the part is turned on centers, it is best if they remain in the part. This may facilitate inspection or rework of the part, and the time to remove them and the material lost are avoided.

The tolerances that are reasonable for lathe work in production are ±0.001 for diameters to 1 in. and ±0.003 for diameters to 4 in. There is a slight increase with increasing diameter. Greater tolerance is desirable, and smaller tolerance can be achieved at a considerable increase in cost. The minimum tolerance desired on length is ±0.005, but ±0.002 can be maintained on short lengths. The concentricity between diameters machined in the same setup can be held to 0.004 total indicator reading (TIR) in production. This means that the actual centers of the diameters fail to coincide by not more than 0.002 in. Diameters not machined in the same setup must have greater tolerance and will cost more. The minimum surface finish practical in production is 63 μin. Drilled holes are reasonably true to a depth equal to 5 times their diameter.

A turret lathe is the same as an engine lathe, except for the provision for mounting several different cutting tools in such a manner that they can quickly be put into operation. Thus, the turret lathe can be used for the production of parts where the quantity does not justify using an automatic screw machine, or where the size of the parts is beyond the capacity of the automatic screw machines. The tolerances given for an engine lathe also apply to a turret lathe.

An automatic screw machine is a form of lathe where all of the operations are performed automatically, including the feeding of the stock from which the parts are made. It is intended for large-quantity production and is seldom used where less than 1000 parts are to be produced in the same lot. The maximum-size part that can be made on the machines most often available is 2.75 in. diameter by 6 in. long.

If a part is to be made on an automatic screw machine, it should be designed so that as many of the surfaces as possible can be machined before the part is cut off. If the part consisted of outside diameters that differed greatly in size, it could probably be more economically produced by hot or cold heading and then machining, because of the smaller amount of machining required.

Tolerances of ±0.010 on both diameters and lengths is desirable. However, where necessary, ±0.003 can be held in production. A concentricity of 0.005 TIR is usually maintained. For external surfaces that are machined simultaneously, or for shallow holes, 0.001 TIR can be achieved with care. Angles of ±1° can be held in production.

A Swiss automatic screw machine is preferred for small or slender parts.

To be economical, a large number of parts must be produced. The maximum diameter is 0.50 in., and lengths less than 2.75 in. are preferred, though parts 9 in. long can be produced. Centerless ground stock should be used where close tolerances are required.

A tolerance on diameters of ±0.0005 can be held in production, and with care on the smaller machines ±0.0002. A length tolerance of ±0.003 is desirable, though considerably less can be maintained. A concentricity of 0.001 TIR is held in production. A surface roughness of 16 μin. can be maintained.

Boring is a process whereby an existing hole is made larger. It can be done on many machines. However, when a machine other than a boring machine is used, the quality of the work generally suffers. Boring is also used when the size of the hole is larger than can be accommodated by other processes. It costs about twice as much as drilling. Drilling is limited to about 3 in. in diameter, but boring is almost unlimited.

The tolerances that may be held with a boring machine in production are ±0.0005 for a 1-in. diameter. For smaller diameters the tolerances may be closer, and they should be increased slightly with larger diameters. Tolerances half as large can be maintained where necessary. Depths to ±0.002 may be held.

A jig borer can be used to produce parts to extremely small tolerances on both the size and location of the holes. Though these machines are used mainly in toolmaking, they may be used for production if the cost can be justified.

4-5 MILLING

Milling is a process that uses cutters which have many teeth. It is not as well suited to quantity production as some of the other machining processes, such as an automatic screw machine. Because of the great variety of cutters available, it is a very versatile process. However, the cutters have a great influence on both the configuration and size of many of the features of a part produced by milling. A brief description of the more common cutters may be in order. A plain cutter, shown in Fig. 4-15, has cutting edges only on the periphery. A side cutter has cutting edges on the periphery and both sides. A half side cutter is the same as a side cutter except that it has cutting edges on only one side. Standard cutters are available that will make the cuts illustrated in Fig. 4-16; for Fig. 4-16a the angle can be 45° or 60°, and for Fig. 4-16b it can be 45°, 60°, or 90°.

Standard side cutters are available in widths from $\frac{3}{16}$ to $\frac{1}{2}$ in increments of $\frac{1}{16}$, and from $\frac{1}{2}$ to 1 in. in increments of $\frac{1}{8}$. Slitting saws, which are similar to side cutters, are available from $\frac{1}{16}$ to $\frac{3}{16}$ in increments of $\frac{1}{32}$. Plain cutters are available in all the widths specified for side cutters and, in

Fig. 4-15 Plain milling cutters. (*National Twist Drill & Tool Co.*)

addition, greater widths to 12 in. The standard diameters are 2, 2.5, 3, 4, 5, 6, 7, and 8 in. Not all the widths are standard for all diameters, but the narrower widths are standard in smaller diameters.

A part to be produced by milling must be rigid enough to withstand the cutting force without deflection. Thought should also be given to holding the part in the machine. Holes that are part of the design may be used to clamp the part, in some instances, or holes may have to be added for this purpose. Sometimes it is necessary to add lugs, the only purpose of which is to provide for properly holding the part. If holes are used, there should be at least two of not less than 0.25-in. diameter and located as far apart as possible. The part should be designed so that the maximum number of surfaces can be machined in one pass. It should be kept in mind that removing the part and placing it in another position will take more time, require greater tolerance, and allow another opportunity for error.

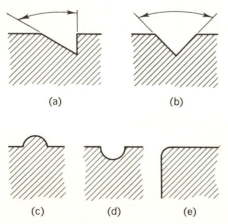

Fig. 4-16 Shapes produced with milling cutters.

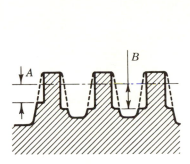

Fig. 4-17 Gang milling.

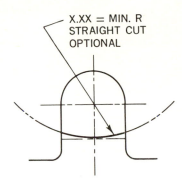

X.XX = MIN. R
STRAIGHT CUT
OPTIONAL

Fig. 4-18 Arrangement for milling.

A slot to be produced by a side cutter should have a nominal size equal to the nominal width of the cutter, with a plus and minus tolerance. The mating tongue can be easily machined to the thickness required to provide the desired tolerance by mounting two half side cutters with a spacer of the proper thickness. Slot depth in excess of 3 times the cutter width should be avoided.

Advantage should be taken of gang milling wherever possible. This is placing several cutters on the same arbor, with appropriate spacers, so that several surfaces are milled in the same operation. This is illustrated in Fig. 4-17, where six surfaces are machined at once. When designing for gang milling, the dimensions A and B should be such that cutters of standard diameter can be used. The part should be designed so that a note similar to that shown in Fig. 4-18 may be used.

The tolerance desired for milling, either the width of a slot made by the cutter or the thickness of a lug, is $+0.007/-0.003$. However, with care, a tolerance of $+0.003/-0.001$ can be achieved. The surface roughness usually produced is between 63 and 250 μin. With care, and especially with carbide cutters, the finish can be considerably improved.

Another type of cutter used in milling is the end mill, one form of which

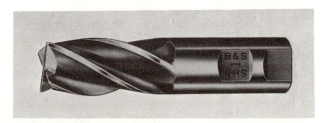

Fig. 4-19 End mill. (*Brown & Sharp Mfg. Co., Cutting Tool Div.*)

is shown in Fig. 4-19. Compared with a regular milling cutter, an end mill is frail and is thus less accurate and much slower in removing metal. End mills should be avoided, but when they are used they should be as large as possible. They are available in diameters from ⅛ to 2 in. The minimum desired size for production is ⅝.

All internal radii should be the same size on parts that are to be produced with an end mill. Also, all fillet radii should be the same size. This is to permit a single cutter to be used. The fillet radius is produced by the radius on the corner of the cutter and should not be too large, for this interferes with the cutting action. However, it should not be less than 0.05. The desired radius is 0.06, with a ±0.01 tolerance. Where a large fillet radius is required, a ball end mill can be used, which has the fillet radius equal to one-half the diameter of the cutter. Parts should be so designed that the portion produced by the bottom of the end mill is at a 90° angle to that produced by the side, as shown in Fig. 4-20a.

Slots should be made with a regular milling cutter, as shown in Fig. 4-21a, rather than with an end mill, as shown in Fig. 4-21b, because such

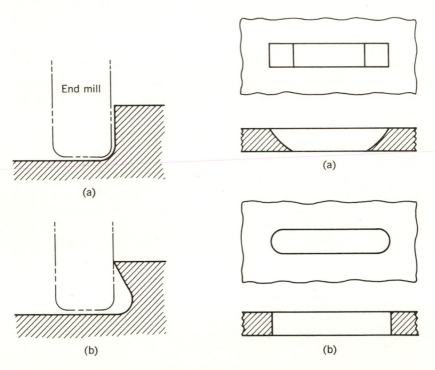

Fig. 4-20 (a) Preferred configuration when using an end mill; (b) unsatisfactory configuration.

Fig. 4-21 (a) Preferred slot; (b) unsatisfactory slot.

slots are more accurate and cheaper. Several keyways spaced radially on a shaft should be avoided, because close angular tolerances are difficult to maintain. A regular milling cutter, rather than an end mill, should be used to produce long keyways. This is to facilitate production and to reduce the stress concentration. Stress concentrations will be discussed in Chap. 7.

4-6 DRILLING AND REAMING

Drilling is used to produce holes to final size, also to prepare for other processes such as reaming, boring, and tapping. Drilling can be accomplished with many different types of machines, and when the holes are not too large, a portable unit can be brought to the work.

Standard drills should be used. Because of the great number of standard sizes, this is not a limitation of any consequence. Fractional sizes have diameters from $\frac{1}{64}$ to $1\frac{3}{4}$ in increments of $\frac{1}{64}$, from $1\frac{3}{4}$ to $2\frac{1}{4}$ in increments of $\frac{1}{32}$, and from $2\frac{1}{4}$ to $3\frac{1}{2}$ in increments of $\frac{1}{16}$. Number sizes have diameters from 0.0135 for a No. 80 to 0.228 for a No. 1. The increment for some of the smaller sizes is 0.001. Letter sizes have diameters from 0.234 for an A drill to 0.413 for a Z drill. Other sizes may be obtained. Diameters in increments of $\frac{1}{32}$ should be used for drills larger than 1 in.

When using ordinary drills and drilling methods, depth equal to 5 times the diameter of the drill is a desirable limit. Where more than a single drill can be used in producing the hole, the limit can be doubled. Where special drills and methods are used, holes 50 times the diameter of the drill can be produced.

In operation, the point of the drill tends to deviate from a straight line. The deeper the hole, the greater the total amount of deviation. This is called runout. Also, a small drill will have greater runout than a larger one for a given depth. An idea of the magnitude of the runout can be gained from the following average values: 1 in. depth — $\frac{1}{4}$ drill 0.006, $\frac{1}{2}$ drill 0.001; 2 in. depth — $\frac{1}{4}$ drill 0.024, $\frac{1}{2}$ drill 0.005.

Holes should be drilled through. But where blind holes are used, the bottom of the hole should have the form of the drill point, which is generally a 118° included angle. Also, the minimum tolerance on the depth should be $+0.06/-0.00$.

Some points of good design practice pertaining to drilling are as follows: A drill should be started on a surface perpendicular to the axis of the hole to be drilled. Situations where a portion of the hole breaks through into an opening, as well as an odd number of holes in a circular hole pattern, should be avoided. Allowance must be made for the drilling machine that drives the drill. Counterdrilled holes, that is, holes that consist of two or more diameters, should allow the bottom of each of the holes to have the form of the drill point.

Drills tend to cut oversize, and appropriate plus tolerances for several sizes are: 2-in. diameter — 0.015; 1-in. diameter — 0.010; ¼-in. diameter — 0.007. A minus tolerance is desirable, to allow for worn drills; from one-thousandth for small drills to several thousandths for large drills is appropriate. With care, the plus tolerances given can be reduced to almost one-half, and the minus tolerance made zero.

Reaming is a process whereby the diameter of a hole is made more accurate and the surface finish is improved. It cannot correct the location or alignment of the hole. The usual procedure in design is to use a hole made by a standard reamer, and to make the mating part smaller by the amount necessary to provide the desired fit. It is also a good practice to avoid blind holes. However, where this is not feasible, the design should be arranged to require the hole to be reamed only to within 0.12 of the bottom. This is to allow for the chamfer on the reamer and the chips that will collect in the bottom of the hole while reaming.

Standard reamer diameters are: ⅛ to ¾ in. in increments of ¹⁄₆₄, ¾ to 1 in. in increments of ¹⁄₃₂, 1 to 3 in. in increments of ¹⁄₁₆. The minimum desired tolerance is ±0.001 for diameters less than 1¼, and +0.0025/−0.0015 for larger holes. With care, these tolerances can be reduced to one-half. The surface roughness produced is generally between 63 and 32, but with care, it can often be reduced to 16. It should be noted that many of the nonferrous metals do not produce the smooth finish when reamed that is possible with ferrous metals. Therefore, the lower values just given may not apply.

Two other processes performed with drilling machines that deserve mention at this time are counterboring and spotfacing. Both are performed with a counterboring cutter. Counterboring is enlarging a drilled hole considerably for a short distance and produces a flat bottom. Counterdrilling, which is producing the enlarged diameter with a drill, is more economical than counterboring. The fillet radius at the bottom of a counterbored hole should be 0.06 ± 0.01.

Spotfacing provides a circular area around a hole that is at a right angle to the axis of the hole. It is used primarily as a seat for bolt heads and nuts. The diameter should be slightly greater than the distance across the corners of the bolt or nut that will be placed against it. A spotface may consist of a recess, or a boss may be provided to be machined by spotfacing. The boss is the better practice, and the height after spotfacing should be greater than 0.03. The diameter of the spotfacing cutter should be greater than the diameter of the boss. Back spotfacing, that is, pulling the cutter against the surface, can be accomplished, but should be avoided, because it costs over twice as much as spotfacing.

The diameter of standard counterboring cutters is as follows: ¼ to 3 in. in ¹⁄₁₆ increments, and 3 to 4¾ in. in ⅛ increments. Back spotfacers are available from ½ to 2 in. in ¹⁄₁₆ increments. The tolerances that may be held are the same as for drilling.

4-7 BROACHING

The two most desirable characteristics of broaching are accuracy and high production rate. It is unsurpassed for producing internal keyways, splines, and serrations, and it is the most practical method of producing internal spiral splines.

A properly made broach of high-speed steel can be expected to produce 75,000 parts to the same degree of accuracy as the first parts. Because of the large number of cutting teeth, little heat is generated and thus there is little chance of distortion. Another advantage is that neither the configuration of the part nor the feed of the tool is dependent on the operator; thus, there is very little scrap associated with broaching.

Because of the high cost of a broach, the number of parts to be manufactured is a very important factor in determining the feasibility of using the process. The number of parts required to make the process a practical choice varies greatly, but it is seldom used for less than 2000 parts, and if there are other appropriate methods of performing the operation the number may be over 25,000.

Any material that can be machined by the usual methods can be broached. The hardness should be Rockwell C25 to C35. Softer metal can be handled successfully, but there is a tendency for it to tear and for the metal to adhere to the broach, thus degrading the accuracy and surface finish. Harder metals can also be broached, but they cause excessive wear of the broach.

One of the principal limitations of broaching is that a chip the length of the broached surface is removed by the cutting edge and must be retained in the space between the teeth for the entire pass across the surface. This means that the length of the broach depends on the depth of the cut; that is, deep splines would require a longer broach than would shallow serrations. It also means that a long spline requires a longer broach than does a short spline. If the desired configuration cannot be achieved with a single broach, additional broaches may be used. But this, of course, increases the tooling cost and the time for performing the operation.

There must be at least two teeth in the work at all times, but for thin parts the teeth can be placed at a slight angle to satisfy this requirement. Very thin parts can be broached if they are designed to permit their being stacked so that there can be three teeth in the stack at all times.

The characteristic tolerance is ± 0.001, though a tolerance of ± 0.0003 can be held, but at a considerable increase in cost. This does not apply to round or square holes, which can be easily held to ± 0.0005 in production. Broaching is very consistent; the variation in size of holes about 1.500 in. in diameter is less than 0.0002 from one part to the next. A surface finish of 32 can be obtained consistently in production, provided there is no variation in hardness of the material being broached.

Because of the large number of teeth cutting at any one time, the cutting force is comparatively large; it may be greater than any load imposed on the part in operation and probably in a direction different from the maximum load in operation. Consideration of this fact should not be overlooked when designing a part that will likely be produced by broaching.

Whenever a square hole is required in a design, rather than a round hole, it is because the corners serve some purpose; however, there is usually no reason why the hole cannot be made as shown in Fig. 4-22, rather than a perfect square. This shape is preferred, because there is less metal to remove, it provides a channel for the cutting fluid, and the corners facilitate manufacture and maintenance of the broach. Often, a long hole can be redesigned, as shown in Fig. 4-23, which will greatly facilitate broaching. The center portion can be produced by boring, or if the part is a casting, by coring.

In broaching a spline, the usual procedure is to machine the hole to the inside diameter of the spline, then to cut the spline with a broach. But if the concentricity between the inside diameter of the spline and another part of the spline must be held to a very small tolerance, the broach can be made to produce the final inside diameter. It is appropriate to point out that a spline seldom needs to be longer than its pitch diameter.

Spiral splines can be readily formed by broaching, if the helix angle is less than 60°, by use of a spiral lead bar in the machine. However, it is more economical if the helix angle is less than 15°, for the broach will then generate the proper helix itself when pulled with a special head that permits rotation.

In order to reduce the cost of a broach, or to bring the operation within the capacity of a certain machine, multiple keyways, even splines or gear teeth, can be broached one at a time or a few at a time and the part indexed

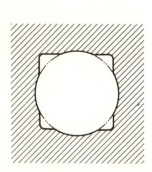

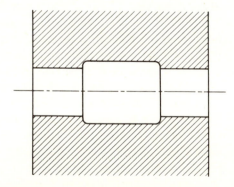

Fig. 4-22 A square hole produced by broaching.

Fig. 4-23 Arrangement for a long broached hole.

between passes. Of course, this is much slower. A broaching machine is not essential to the process of broaching. A fixture with an air-operated actuating cylinder could be used to advantage in making small parts, especially with the softer materials such as brass and aluminum.

4-8 OTHER MACHINING PROCESSES

A planer is used to machine flat surfaces on large, bulky parts, such as machine frames. A 20-ft stroke is not uncommon. Planing is not suited to large-quantity production. In designing a part that will be machined on a planer, thought should be given to locating and clamping the work to the table. Providing pads, brackets, and holes just for this purpose is appropriate. The part should also be designed for a minimum of deflection when subjected to the handling, clamping, and cutting forces. The additional material to accomplish this is justified. As many as possible of the surfaces to be machined should be in the same plane. And if the tool must be stopped at a certain point in its travel, because a portion of the part is above the plane being machined, a relief should be provided. The relief is usually cast and must be lower than the finished surface. A width of several inches would be appropriate for a large part. Tolerances that may be expected depend, to an extent, on the size of the work, but about ± 0.005 is a good average, and for cast iron it can be somewhat closer. The surface roughness averages about 200. And, again, cast iron will permit an improvement.

Grinding makes possible the manufacture of parts from materials that cannot otherwise be machined. Grinding can also be used in the production of parts to tolerances and surface roughness difficult to achieve by other methods. Grinders that hold parts on centers are made that can produce parts 3 ft in diameter and 16 ft long. The length of a part that can be made in a centerless grinder that can produce a profiled part is limited to the width of the grinding wheel, but parts 3 ft long can be accommodated.

When a part is to be ground on centers, the centers should remain in the part at least until all manufacturing processes have been completed. It is best if the parts are balanced about the axis on which they will turn when ground. This is to avoid the necessity of providing a fixture to prevent difficulties that would result from vibration.

Wherever possible, a relief should be provided, so that a radius need not be ground. The relief should have the form indicated in Fig. 4-24. The reliefs apply to either internal or external surfaces. The minimum width should be 0.06. Where the fillet must be ground, it should be large, and all the ground fillets on the part should be the same size.

Surface grinders produce flat surfaces, which should be in the same plane. Where this is not feasible, the surfaces should be arranged as shown in Fig. 4-25a, rather than as in Fig. 4-25b. The use of reliefs is appropriate,

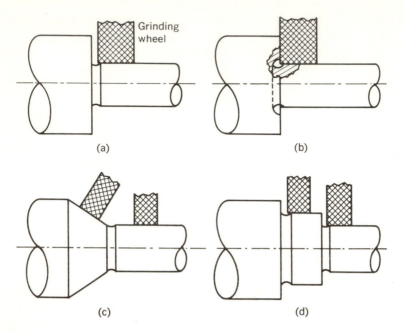

Fig. 4-24 Grinding reliefs.

and the area to be ground should be limited as shown in Fig. 4-26*a*, rather than as shown in Fig. 4-26*b*.

The tolerance on a diameter of moderate size, either internal or external, that can be held in production is ±0.0001, though ±0.0005 would cost considerably less. The tolerance between shoulders can be held to ±0.0003. Flat surfaces can be held to a flatness tolerance of 0.0002, parallelism of 0.0004, and length from ±0.0005 to ±0.002, depending on the size. Again, larger tolerances should be allowed where possible. The average surface roughness produced by finish grinding is 16. With care, smoother surfaces can be produced, especially with ferrous metals.

Honing is a finishing process that can correct a hole that is slightly out of round or tapered, but it cannot correct position or alignment. It is a quantity-production process, and the size of a part is usually not a limitation. Reliefs should be provided, and at the bottom of a blind hole they should be over 0.20 wide. Bores that must be aligned should not be blind. Ferrous and most nonferrous metals may be honed, but very soft metals, such as babbitt, are usually not honed, because they clog the pores of the stones that are used. Tolerances that can be maintained in production are from ±0.0001 to ±0.0003, depending on size. A surface roughness of about 8 can be produced with ferrous metals and bronze in production, and with care, a smoother surface can be obtained.

Lapping is a surface-refining process, and as a maximum of 0.0002 is

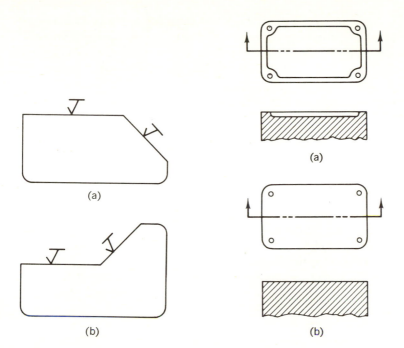

Fig. 4-25 (a) Preferred configuration for surface grinding; (b) unsatisfactory configuration.

Fig. 4-26 Arrangement of large flat surfaces.

removed, it cannot correct any defects in the hole. Generally, honing is performed to prepare for lapping. It is more limited with respect to size of parts than is honing. With hard steel, lapping can produce as smooth a finish and as close a fit between two parts as is desired. When two parts are lapped to a very close fit, they are not interchangeable and must be identified as a pair.

Superfinishing is also a surface-refining process. It is suited to quantity production, because of the short time required to produce the required finish even from a rather rough surface. Reliefs are desirable, and though grooves and slots cause difficulties they can be accommodated. A surface roughness of about 2 is produced in production. Tolerance depends on the tolerance before superfinishing and the required surface roughness, because about 0.00005 in. is removed from a surface for each 10 μin. reduction in surface roughness.

4-9 SHEET AND PLATE PROCESSES

Of the many means by which sheet and plate are made into finished or semifinished parts, stamping is the most often used. It is also one of the most widely used of all manufacturing processes for quantity production. Producing parts by stamping at the rate of 1000 per minute is not unusual. Stampings are generally made of metal from a few thousandths thick to ⅜ thick, though twice this thickness can be stamped cold, and if heated the thickness may be over 3 in.

About 10,000 parts are generally required to justify the cost of a stamping die, though a smaller number of simple parts can be economically produced with "short-run" dies.

Steel is the most common material used for stampings, and a carbon content less than 0.20 percent is preferred. Hot-rolled steel is used where finish is not a factor and cold-rolled where the parts will be plated or painted.

When designing parts to be produced by stamping, the fact that they must be of uniform thickness should be kept in mind. It should also be remembered that stampings can be joined by furnace brazing to produce a part that cannot be stamped.

To facilitate making the die, stamped parts should have both corner and fillet radii, except where the outline of the part and the edge of the stock coincide; for example, Fig. 4-27a is preferred to Fig. 4-27b. The radius for a part similar to that shown in Fig. 4-28 should be considerably greater than one-half the stock width.

Holes that are punched at the same time the part is stamped cost much less than drilling later. But there are limitations: they should not have a diameter less than the thickness of the metal; the minimum distance between holes, or between a hole and the edge of the metal, should be twice the metal thickness but not less than 0.125. One of the disadvantages of sheet metal parts is that the holes provide little bearing area or insufficient length of thread for tapped holes. An effective way of overcoming such disadvantages is to extrude the holes rather than punch them. An extruded hole has the

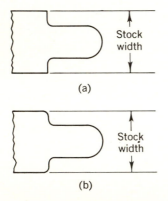

Fig. 4-27 Radii for a stamped part.

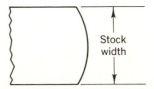

Fig. 4-28 A radius should not equal half the stock width.

Fig. 4-29 Increasing the bearing area in sheet metal.

Fig. 4-30 Boss produced by stamping.

metal pushed out to form a flange around the hole, as shown in Fig. 4-29. The height *h* for moderate-size holes can be one-half the diameter of the hole.

A stamping operation that can often be used to advantage is a small boss formed on one side of the part, as shown in Fig. 4-30. It can be used with a hole in another part to provide alignment. The height should be limited to one-half the thickness of the metal, and the diameter to twice the height.

Making simple bends to produce flanges or ears is referred to as forming. When parts are formed along with other stamping operations, right-angle bends are difficult to hold to a relatively close tolerance, especially with hard or thick metal. To avoid cracking, bends should be made across the grain of the metal. The grain will be parallel to the long edge of the stock. The bend radius, that is, the inside radius around which the metal is bent, should be equal to or greater than the thickness of the metal. The higher-strength steels require a larger radius, often as much as four times the thickness. The height *h* of a flange, as shown in Fig. 4-31, should not be less than 3 times the metal thickness plus the bend radius. Where flanges extend over only a portion of the part, a bend relief should be provided, as shown in Fig. 4-32. The minimum distance *a* between a hole and a flange, as shown in Fig. 4-33, is the hole radius, plus twice the thickness of the metal, plus the bend radius.

The tolerance desired on dimensions that should be held reasonably

Fig. 4-31 Flange height.

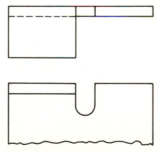

Fig. 4-32 Bend relief.

close is ±0.010. Half this can be held where necessary, and with care, critical dimensions such as hole locations can often be held to ±0.001. The tolerance desired on hole diameters of moderate size is ±0.005, though ±0.0005 can sometimes be held, but the cost will be much greater.

Parts such as kitchen kettles or automobile fenders are formed by a process known as drawing or deep drawing. It is basically a stretching process. The sheet is stretched, or drawn, to the desired configuration in appropriately shaped dies. Metal as thick as ¾ in. can be drawn cold, and if heated the thickness can be over 3 in. A special deep-drawing steel with a carbon content of 0.08 percent is preferred where the drawing is severe. Parts with round cross sections are much easier to produce than box-shaped parts. Cup-shaped parts with flat bottoms are preferred to those with spherical bottoms, but the radius at the bottom should be large relative to the metal thickness. Complex parts often require considerable development to make the dies, and changes in design are usually required. Large tolerances should be allowed on parts of complex contour. A tolerance for small simple parts of ±0.005 is reasonable for production.

Roll forming is the process by which sheet metal is formed into cross sections of various shapes by passing the sheet through a series of rolls that progressively form the desired shape. Figure 4-34 shows a few of the many shapes that can be produced. These sections are generally used in fabricating structures by riveting or welding. A common application is airplane framework. Metal from a few thousandths to ¾ in. can be used, and the length is limited only by the length of the stock, which for coiled strip is practically unlimited. The minimum length is about 2 ft. A single machine can produce about one mile of section in an hour, and several million feet can be produced before the rolls need maintenance. Thus, it is obvious that this is a large-quantity process, and quantities of this magnitude are necessary to justify the cost of making rolls. However, some simple sections can be produced in a range of sizes by adjustable rolls. Large corner radii should

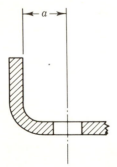

Fig. 4-33   Relation of a hole to a flange.

Fig. 4-34 Sections made by roll forming. (*Yoder Co.*)

be avoided, and the bend radii should not be less than the thickness of the metal for those metals easily formed. For tempered metals, the minimum may be 5 times the thickness. When a rather long, flat surface is adjacent to an edge, it should be formed in some manner rather than left as a raw edge. The tolerance between the elements of the cross section can be held to ±0.005 on moderate-size sections.

Brake forming is a process for producing cross sections similar to those made by roll forming. There are great differences in the processes, however. The length is limited by the length of the brake and its capacity. The cost of tooling is very low; therefore, the process is suited to the production of small quantities. The maximum thickness that can be accommodated by brakes generally available is 1 in., and the length is limited to 30 ft. But when the thicker metal is used, the length must be reduced. The minimum bend radius is equal to the metal thickness, and a larger radius is desirable. Of course, several bends requiring different dies could be used to produce a section. Holes can be punched with a brake. The desired minimum spacing is ½ in. for moderate material thickness. The desired tolerance on location is ±0.06, though for short lengths and small sections one-half this can be held. The tolerances on the diameter and spacing of holes should be ±0.005, though diameters can be held closer.

4-10 METAL-TREATING PROCESSES

Many metal-treating processes require little more than the decision to use them; however, others require provisions in the detail design of the parts to make their use feasible or economical. Heat-treating is such a process. Parts can be made from a metal that can be hardened throughout, or various processes can be used to produce a hard surface on metals that cannot be through-hardened. It is better for a designer to specify what is required of the heat-treatment than to dictate the process to be used. Of course, he must know that what he specifies is feasible. It is a poor practice to specify both the strength and the hardness required of a heat-treating process.

Often the hardness of a part is dictated only by the need to resist wear; at other times, both toughness and hardness are required. In such situations, case hardening is indicated. The depth of the case required depends on the amount of wear that must be allowed for in the life of the part and the amount of distortion in the part that will have to be corrected by grinding.

Carburizing is the process in which carbon is added to the surface and then the parts are quenched. It is the most widely used of the case-hardening processes. A case thickness of less than 0.02 is used for wear resistance where loads are low, about 0.03 is used for moderate loads, and about 0.07 for high loads and severe wear. Carburized cases will lose hardness at high temperatures.

In carbonitriding, both carbon and nitrogen are added to the surface. There are two methods: cyaniding, which produces a case of about 0.01, and the gas method, in which the case is about 0.03. Nitriding uses gas to add nitrogen to the surface. It is a long process, and special steels are generally used. The case is usually less than 0.015, though thicker cases can be produced. The case will not soften at high temperature.

Parts that are to be heat-treated should be as symmetrical as possible, with any holes in a balanced arrangement. A considerable difference in the thickness or width should be avoided. Changes in section should be made gradually. Corner radii, thread reliefs, and generous fillets are appropriate. It is best to avoid having a shoulder and depth coincide, as shown in Fig. 4-35.

It is occasionally desirable to have a portion of a case-hardened part remain unhardened. There are several methods of accomplishing this, but all involve extra cost.

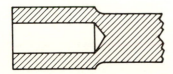

Fig. 4-35 An unsatisfactory arrangement for heat-treating.

Plating is frequently used. The thickness required depends on the function that the plating is to perform and the conditions of service. For decorative applications, the thickness may be less than 0.001, but for functional applications, a thickness of 0.010 would be closer to the average.

Generous corner and fillet radii should be provided. Parts in which air could be trapped when placed in the plating solution require an air hole; otherwise, there will be no plate in the area of the trapped air. Parts that could trap the plating solution when removed should have a drain hole, because the solution is expensive, and a part that causes it to be wasted will cost more to plate. Deeply recessed parts will require special methods to plate the inside, which increases cost. Where appearance is a factor, large flat surfaces should be slightly crowned.

Where tolerances after hard chrome plating must be closely held, grinding must be used, as the plating cannot be machined. If larger tolerances are permissible, parts can sometimes be plated to size. It must be remembered that, though chrome plate is hard, if the part is subjected to a compression load that causes excessive deformation of the metal supporting the plating, it will peel off.

Shot peening is a surface treatment often performed on highly stressed parts which will be subjected to many applications of the load. Often, parts are peened only on selected surfaces. A rubber tape is used as a mask. Surfaces that are to be peened must be easily reached with the stream of shot. Sharp corners should be avoided. The effect on tolerances is almost negligible, but honing or lapping may be used after peening where tolerances require it.

4-11 PROCESSING PLASTICS

Parts can usually be made from plastic by machining stock. The capabilities and limitations are similar to those encountered in machining metal. However, when a large quantity of parts of medium size are to be made from plastic, molding is usually the more economical process. The molding process used is similar to die casting and is subject to the same limitations. The wall thickness must be as uniform as possible. The minimum thickness varies with the kind of plastic and the method of molding, but the preferred minimum thickness is usually 0.06. The draft required also varies with the process and with the plastic used. The desired minimum is about 3°, but it can be considerably less, especially for external draft. In some instances, as little as 1/8° may be acceptable. Corner and fillet radii should be as large as the function of the part will permit. The desirable minimum is equal to one-fourth the wall thickness. Thick sections should be avoided, as for casting, and any difference in thickness should be made gradually. The edge of a hole should not be closer to the edge of another hole or to the edge of the part than a distance equal to the diameter of the hole. Blind holes are limited to about 3 times their diameter; through holes can be considerably deeper. It is better to drill small holes. Inserts can be used to advantage in some applications, even though they increase the cost considerably. Sharp corners on the inserts should be avoided. Some method of securing them, such as knurling, must be provided, and they should have a smooth surface where they protrude from the molded part.

The minimum tolerances for a small part are: between points in the same side of the mold, ±0.003; across the parting plane, an additional ±0.010 is required. It should be remembered that dimensional changes with changes in environment are relatively large, and that the tolerances for parts made of plastic are not as critical as for metal parts, because of their greater resiliency.

4-12 MISCELLANEOUS PROCESSES

Powder-metal parts are made by compressing a metal powder to the desired form and then sintering it to weld all the metal particles together.

Fig. 4-36 Powder-metal parts. (*Chicago Powdered Metal Products Co.*)

The process can be used to produce parts that can be made in no other way, such as porous bearings and filters. The size of a part seldom exceeds 1 ft in diameter. In instances where other processes can produce the part, the number required to justify using powder metal is largely dependent on the number and complexity of the machining operations that can be eliminated, and thus may be 5000 or 25,000.

The process is severely limited with respect to the configurations of the parts that can be produced. Only parts similar to those shown in Fig. 4-36, which permit pressing of the powder, can be made. The preferred minimum thickness is about 0.06. Fillet radii should be provided, but should be neither too small nor too large. Approximately 0.015 for a moderate-size part is appropriate. A chamfer of about 0.015 × 45° should be used in place of a corner radius. The tolerance for small simple parts perpendicular to the direction of pressing is about ±0.002. In the direction of pressing a much greater tolerance is required.

Metal spraying, or metallizing as it is sometimes called, is a process by which molten metal is sprayed in much the same manner as paint is sprayed. The outstanding feature of this process is that it permits a part to be made of a metal that satisfies most requirements in addition to having the properties and characteristics of another metal in those portions of the part where they would be more desirable.

A great many metals can be sprayed, and the thickness can be from only a few thousandths to as thick as desired. The hardness and wear resistance of a metal when sprayed are greater than the metal normally exhibits. Bearing metals have a porosity which allows the absorption of lubricant that improves their qualities as a bearing. Parts must be so designed that the sprayed metal does not come to a sharp edge and is not subjected to an intense compression load.

There are a number of processes for removing or shaping metal that are a radical departure from the conventional methods that have been discussed. They permit the use of metals that are difficult or impossible to process by the usual means, and in some instances they make it possible to produce a larger part or to manufacture a part more economically.

In the process called chemical milling, metal is removed by dissolving in a chemical solution. The configuration of the part is produced by masking selected areas and by varying the time that various portions are in the solution. In the process referred to as electrical discharge machining (EDM) metal is removed by a great many small sparks that are caused to jump between the tool and the part. The shape of the tool determines the shape of the "cut," though there is no physical contact between the tool and the part. Electrochemical machining (ECM) is essentially the reverse of electroplating, for metal is removed by electrolytic erosion. Again, the tool determines the shape of the cut but does not come in contact with the part. Hard or tough metals are easily handled, heat is not generated, and a very smooth surface can be produced. ECM is a faster process than EDM.

Sheet metal and plate are formed by creating a powerful shock wave in water. The configuration of the part is produced by the shape of the die into which the metal is forced by the shock wave. The shock wave is generated by causing an enormous spark in the water, or by detonating an explosive. Because the die can be made from an easily worked material and the mating part is eliminated, this process can be used for the production of a few parts.

QUESTIONS

4-1   Why is a designer concerned with manufacturing processes?

4-2   What is the difference between external and internal draft?

4-3   On a casting, where is a corner radius undesirable?

4-4   What is the preferred relationship between the radius of a boss and the diameter of the spotfacer used to machine it?

4-5   Discuss the factors responsible for the relatively large tolerances that castings require.

4-6   What are two of the most important advantages of sand casting?

4-7   What are two of the most important features of die casting?

4-8   Discuss inserts relative to both sand casting and die casting.

4-9   What is the unique feature of plaster-mold casting?

4-10 For what type of part is shell-mold casting most appropriate?

4-11 For a forging, what is the proper relationship between the parting plane and the web?

4-12 Why is a large fillet radius desirable for a forging?

4-13 Why is a corner radius desirable for a forging?

4-14 In the design of a forging, why should a thin web be avoided?

4-15 Discuss the various tolerances that apply to forgings.

4-16 Compare the features of hot upsetting and cold heading.

4-17 What type of part is most appropriately machined on a Swiss automatic screw machine?

4-18 Discuss the effect that milling cutters may have on the design of parts to be made by milling.

4-19 Discuss several points to be considered when designing a part that requires drilling.

4-20 What are several points to remember when designing parts that will be reamed?

4-21 Compare counterdrilling and counterboring.

4-22 What is the primary purpose of spotfacing?

4-23 What are some of the advantages and limitations of broaching?

4-24 Discuss reliefs relative to cylindrical grinding.

4-25 What can be done to increase the length of thread for a part made by stamping?

4-26 Should bends in sheet metal parts be made parallel to or across the grain of the metal?

4-27 Compare roll forming and brake forming.

4-28 What are the more important points to keep in mind when designing a part that is to be heat-treated?

4-29 Why are holes in plated parts sometimes provided just to facilitate plating?

4-30 Discuss inserts relative to molded plastic parts.

4-31 Discuss the possibilities of metal spraying in mechanical design.

SELECTED REFERENCE

"Metals Engineering-Processes," ASME, New York, 1958.

5 DESIGN CONSIDERATIONS

5-1 INTRODUCTION

A designer must continuously make decisions that require sound judgment. On analysis, this judgment is found to consist largely of the careful consideration of a number of relevant factors, to each of which is assigned a degree of importance that is appropriate for the decision being made. This chapter is intended to aid in developing the background which will serve as the basis for such judgment. Thus, a rather wide range of subjects will be discussed, some of which may appear to have little relation to design. However, they are typical of the relevant factors referred to above. Some items will be treated in detail at other places in the text, but consideration of them in general is appropriate here. Consideration of all of these items, of course, will not be required for every design undertaken, but many are involved in even a relatively simple situation.

5-2 FUNCTION

A great deal more is required than proper function to make a product successful; however, if it does not function properly, it will be a failure. It would be appropriate to mention some of the basic concepts that apply to function which do not appear in other sections.

In working out a design, calculations must be made to determine the capacities of components and the size of some of the parts. It must be remembered that many of these calculations are based on assumptions and thus should not be regarded as absolute. Whenever experience or judgment appears to disagree with the result of calculation, the entire calculation should be repeated and then checked by some other method, even though this other method is known to be less accurate. Every effort should be made to bring about agreement. Nevertheless, the final decision must be based on judgment.

There is a difference between what is mechanically possible and what is economically feasible, and this difference is often the difference between success and failure of the design. A product of simple design is almost always cheaper to manufacture. However, designing it usually takes longer and requires greater ability.

When providing for strength, an effort should be made to keep to a minimum those situations involving complex stresses, such as bending, and to arrange the cross sections of members to provide the greatest strength for the amount of material used.

Interferences must be avoided. They may result from failure to make allowance for necessary manufacturing features such as fillet radii and tolerances. An often overlooked source of interference is the deflection of a part when subjected to the maximum load that it is required to take.

When providing for a given function, a designer should ask himself: Is this function really necessary? Is this part required? Could this part serve more than one purpose, or could it be combined with another part? Could this part be less complicated and function just as well?

5-3 STAYING WITHIN CONSTRAINTS

Staying within constraints includes not exceeding a specified size, weight, or volume. It is usually thought that increasing the size will increase the weight, but there are instances where the reverse is true. For example, the weight of a tube used as a column may be reduced by increasing the diameter.

Both weight and size can usually be reduced, if an increase in cost is acceptable. Occasionally, a figure of so many dollars per pound is specified as a guide in determining whether a lighter, more costly design is justified.

It may be that the size or weight requirement is dictated by convenience in shipping and handling. If this is the case, the requirement can be met by so designing the product that it can be shipped in several pieces and readily assembled.

5-4 EFFECT OF ENVIRONMENT

If a product is to be successful in a particular environment, it must be designed with due regard to the effect of that environment. A product designed to operate in an extreme environment will be more costly, and one designed for a more severe environment than it will normally encounter

will be at an economic disadvantage. Thus, a designer must clearly understand the environment which the product is to be designed for.

The most common environment is the normal atmosphere. This can
be more severe than the atmosphere in a gear case which provides a spray
of oil. To prevent damage caused by atmospheric corrosion, attention must
be given to some seemingly unimportant details. Such a detail is the effect
of dissimilar metals in contact. An indication of the severity of the corrosion for two given metals in contact can be obtained from Table 5-1.

TABLE 5-1 Tendency to Corrosion

Silver, graphite, gold
Corrosion-resistant steel
Monel, nickel, inconel
Copper, bronze, brass
Lead, tin
Steel, iron
Cadmium
Aluminum
Zinc
Magnesium

The further apart the two metals are in the list, the greater the corrosion,
and the metal higher on the list will be the least affected.

Because moisture is necessary for corrosion to take place, one method
of avoiding corrosion is to waterproof the area where the two metals are
in contact. Another method is to separate the two with a nonmetallic
washer or gasket. The two metals can also be separated by a nonmetallic
coating. If a metallic coating is used, it should be of a metal that in Table
5-1 appears between the two that are to be joined. For example, a steel
insert used in an aluminum die casting should be cadmium-plated. When
a fastener cannot be of the same metal as that being joined, it should be
one that is higher on the list than the metal joined. If it were lower on the
list, it would corrode to a greater extent than would the other metal, and
because of its relatively small size would soon become ineffective.

Other common environmental factors are:

1.   Dirt — Including mud, sand, and dust. Dirt must be excluded
from gears, threads, and bearings. Dirt could clog passages used for cooling. A layer of dirt could interfere with the capacity of a unit to radiate
heat and thus cause overheating.

2.   Heat — The heat may be of short or long duration. If it is of

short duration, shielding the unit is usually all that is necessary. But for long duration, the unit must incorporate a means of cooling, or be constructed from materials and components not adversely affected by heat.

3. Cold — A problem often caused by cold is the ice that forms. It can cause a mechanism to malfunction or can even cause a structural failure. A mechanism lubricated with oil has much more friction at low temperature.

4. Vibration — The unit can be so designed that the vibration does not cause trouble, or the unit can be protected from vibration by a suitable mounting. Which method to use depends on the sensitivity of the unit and the magnitude of the vibration. In some instances, vibration can be adequately provided for merely by using nuts that incorporate a means of preventing their working loose.

5-5 EFFECT ON ENVIRONMENT

The effect that a product has on its environment is a factor that requires consideration, for if it is undesirable and could have been avoided or reduced, the product will likely be unsuccessful. Noise is a common undesirable effect that a product has on its environment, and one that a modest effort in design can usually reduce.

The law which states that the intensity of a sound is inversely proportional to the square of the distance from the source is valid only when there are no surfaces to reflect the sound.

Sound can be produced by a vibrating surface. Vibration can sometimes be reduced merely by selecting an appropriate machine element; for example, one type of chain is quieter than another, and one type of gear produces less vibration than another. Noise can be reduced by placing an isolator between the source of vibration and the surface that is producing the sound; for example, a motor mounted on a frame to which sheet metal panels are also attached produces much more noise than if the motor were alone on a concrete base. If a resilient mounting were placed between the motor and the frame, or even between the frame and the panels, the noise would be reduced.

Noise can also be reduced by applying sound-deadening material to the panel. A sound barrier can be employed, but it must be remembered that noise can flow through openings in the barrier.

5-6 LIFE

The life of a product is the length of time or the number of times it is operated before it is so worn or deteriorated that it no longer performs properly. There are two kinds of life: operating life is the life with the product operating normally, and shelf life is the life of the product when it is not operating but is deteriorating in storage. Various combinations of operating life and shelf life are required, depending on the product.

Generally, the longer the life that is required, the greater the cost; thus, it is uneconomical to provide a longer life than is necessary.

Factors such as fatigue, wear, and corrosion have a great effect on the life of a product. Some parts of a product will have a shorter life than others, because of these factors. It is uneconomical to have a greater life in some parts than in others. There are two ways of providing for the life of parts that are subject to conditions such as wear, fatigue, and corrosion. The part can be designed to last the total life of the product, which may require expensive materials and processes. The other way is to design the part using inexpensive materials and processes, but in a manner that it is easily replaced.

Designing a product that must have a long life using new materials or processes is complicated by the fact that tests under conditions approaching those that will be encountered in service cannot be made. Great care must be exercised in using data pertaining to life that have been derived from short tests.

5-7 RELIABILITY

A product possessing reliability performs correctly each time it is operated. There are different degrees of reliability, and the higher degrees are generally achieved at a higher cost. Thus, a degree of reliability in excess of that required may not be appropriate. Reliability should be thought of as a property of the product that behaves in accordance with physical laws, the same as the kinematic arrangement, and that is produced in the same manner as the desired motions, that is, by designing it into the product.

A design with a small number of simple parts is inherently more reliable than one with complex parts or with a large number of parts. This is because, with the simple design, there is less chance for something to go wrong. Also, in the interest of reliability, any nonessential functions should be eliminated. The concept of "fewer and simpler" contributing to reliability also applies to the functions to be performed by the operator. It is especially desirable to avoid the necessity of following a precise sequence.

One design may be less sensitive to its environment and to the manufacturing tolerances than another. The less sensitive a design is, the greater is its inherent reliability. A greater life than is required often results from providing the required degree of reliability. This must be kept in mind when evaluating a design on the basis of life. In this case, the excess life is unavoidable.

One way of gaining increased reliability is to use a component that has a capacity several times as great as is required. However, it must be remembered that this method is valid only if the component with the larger capacity has a life and reliability equal to that of the component with the smaller capacity. Sometimes a slight sacrifice can be made in a feature such as accuracy to gain greater reliability. This is referred to as a trade-off.

The following situation is overlooked too often, to the detriment of reliability. A linkage is required to actuate a nonessential item, such as an indicator. There is practically no load involved. The linkage is given little attention, which is thought to be consistent with its importance. However, such a linkage can become disassembled or can fail and jam the major control with which it is associated. Thus, its reliability in not causing trouble rather than its reliability in performing its function dictates the thoroughness to be exercised in its design.

Some of the items discussed in other sections, such as assembly and maintenance, also affect reliability.

5-8 SAFETY

There are several instances in which a product can endanger the person using it:

1. The product does not perform correctly. This is considered under function and reliability. If failure of the product is likely to cause serious injury to the person using it, reliability in this instance becomes more important.

2. The product is not used correctly. The designer cannot do much to ensure the correct use of a product. But he should put forth an effort to minimize the possibility of incorrect use creating a hazard and also to minimize the skill required of the person using the product. This is especially important for products to be used by the general public. The relation between the operator and the machine will be discussed later in this chapter.

3. The product performs perfectly. This is the instance where the designer has the greatest responsibility, and where his efforts will produce the greatest effect. Housings and guards must be provided that will protect the operator and that will not unduly interfere with the efficient use of the product. It should not be possible to bypass the safety device accidentally. Thought should be given to the advisability of incorporating an overload release mechanism and emergency controls. There are instances where an emergency control that will override an overload prevention device is

desirable, such as when the consequences of the overload, even destruction of the unit, are less objectionable than the consequences of not having the unit operate.

The fail-safe concept should be mentioned. A product where this concept is used is so designed that the most likely failures will not create a hazard greater than the inability of the product to perform correctly; that is, even though the product fails, the failure is such that the more desirable of the possible situations is produced. For example, an elevator may be so designed that even though failure makes it impossible to raise or lower the load, the load will not be permitted to fall.

5-9 PROTECTION FROM FOREIGN BODIES

An effort should be made to so design a mechanism that objects that may fall into it will not cause it to jam. This is often not possible, and a protecting cover should be incorporated. A cover may be required to prevent a person from being injured by the mechanism, but a cover suitable for this purpose may not provide protection for the mechanism.

There are instances where a control lever must pass through an opening in the cover, and an object could pass through or become wedged in this opening. To prevent this, a shield can sometimes be attached to the lever,

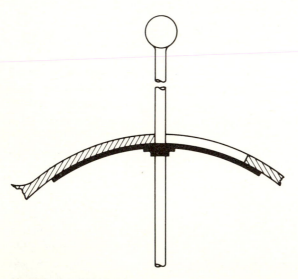

Fig. 5-1 A control lever shield.

which permits the lever to be moved but keeps the opening covered. Such an arrangement is shown in Fig. 5-1. Another method is to use a fabric or rubber cover arranged to permit motion.

5-10 INTERCHANGEABILITY

When assembling a unit, if it does not matter which of several similar parts is used and they can be used without modification, they are interchangeable. Interchangeability is essential if a number of units are to be made economically. Interchangeability is also a characteristic of quantity production, for whenever a large number of parts are made by modern methods they tend to be identical. Thus, designing for interchangeability not only facilitates the process of making the individual parts, but it also facilitates assembly and thus reduces the cost of the product. Operating costs are also reduced by interchangeable parts, for parts likely to wear or break may be kept on hand and, when needed, replaced much more easily than making one to fit.

Interchangeable parts are not identical, because of the variations inherent in all manufacturing processes. As the parts are not identical, the design must provide for the variations; that is, the design must be developed with the idea of interchangeability always in mind.

Interchangeability can be carried so far that it will have a detrimental effect on the design. This is especially true in the case of standard parts; for example, an effort should be made to use a standard ball bearing, but a design should not be compromised in order to use a standard plain bearing.

The degree to which one interchangeable part can differ from another is dependent on the function of the part. A shaft used to support a gear in a speed reducer may have to be held to ± 0.0005 on several diameters to be interchangeable, but the door on a furnace may be interchangeable with a tolerance of ± 0.060.

Interchangeability can be provided for in several ways:

1. Oversized holes can be used in the mating parts to allow for variations in the parts. This method is appropriate where there is little tendency for the parts to become misaligned or where the amount of misalignment is not objectionable.

2. Oversized holes can be used and the mating faces of the parts serrated. This method is appropriate where there is a considerable force tending to cause one part to slide on the other, but can be used only where mislocation equal to the pitch of the serrations is acceptable.

3. Shims may be used to compensate for variations in the parts.

This method will be resisted, in some instances, by those who must approve the design. If it is used, it is essential to so design that if all parts are made to the basic dimensions a shim is required; otherwise, all variations cannot be accommodated.

4. The design can be arranged to use parts that can be adjusted to produce the exact relation to other parts that is desired. This is a more costly method of providing for the unavoidable variations in parts, but it is often possible to combine this adjustment with other required adjustments or with the provision for assembling the parts.

There are many instances where interchangeability of individual parts is essential for manufacturing but inappropriate for repair, though it is desirable to have the assemblies interchangeable. An example is an assembly made by welding several individual parts together. This idea of interchangeable assemblies could be applied to the situation where one method of producing an assembly is employed for making a few units of a new product and another method is used for making a large quantity if production has proved feasible. The assemblies would differ greatly in details but would be interchangeable as assemblies.

5-11 STANDARDIZATION

Standardization, when properly employed, will reduce costs. There are several forms. Standard parts, such as bolts, nuts, and screws, that can be specified by a number should be used wherever possible, because drawings, tooling, and manufacturing paper work are saved. The parts are made in large quantities and thus are probably cheaper than if they were to be made in the company's shop. They are also preferred by those in the assembly and maintenance operations.

Using the same parts for several models of a product will eliminate paper work and tooling. Cost may also be reduced because of the larger quantity.

If the company has standard designs, they should be incorporated where feasible, because less drafting is required and any tooling that may be necessary is already on hand. A designer should also become familiar with any standard processes that the company may have, such as heat-treating or plating, and use them when appropriate. Using such standards eliminates having to specify in detail what is desired, and the production people are more familiar with standard processes.

5-12 ASSEMBLY

Two important reasons for giving considerable thought to assembly are the cost of initially assembling the parts and the cost of assembly when maintenance is performed. The most important reason is often the effect that assembly has on reliability, especially after maintenance or repair.

An effort should be made to so design that the parts cannot be incorrectly assembled. This is particularly important when partial disassembly is required for maintenance or repair. For parts that could be hooked up two ways, the bolts could be made different diameters to permit only the correct way. Care is required to avoid the situation where a part can be installed incorrectly and yet operate correctly under normal conditions, but will not permit the full movement required for complete control, or that will permit overtravel in the wrong direction.

Bolts and screws should not require a closely limited length in order to provide the required clearance, especially where the end cannot be checked. Wherever possible, small parts such as spacers should be avoided, for they may be omitted at assembly. It is important to avoid using simple parts such as washers for a unique and essential purpose.

Thought should be given to the possible advantages of making the assembly in a manner that it cannot be disassembled. Replacing such an assembly may cost less than repairing one that could be disassembled.

A designer must constantly keep in mind that hands and tools are required to accomplish assembly, and that they must move to install fasteners; also, that the assembler should be able to see what he is doing.

5-13 MAINTENANCE

In the design of many products, maintenance is not given the consideration that its importance warrants. Maintenance may have a great effect on both the life and reliability of a product. A designer cannot ensure proper maintenance. Nevertheless, it is his responsibility to so design that proper maintenance is not difficult to accomplish.

The cost of maintenance includes the cost of parts, material, and labor. Another "cost" that is often more important, and that is sometimes overlooked, is the loss resulting from the product not being in usable condition. For example, a bus that is in the shop is not making money for the company.

There are two concepts of maintenance. The first may be thought of as "repair it when necessary," the other as "repair it before failure." The latter is referred to as preventive maintenance. Preventive maintenance is most appropriate where a high degree of reliability is required. It is also desirable in instances where failure would cause considerable damage.

To provide for preventive maintenance, certain parts that are most likely to wear or fail are designed to be readily replaced. It is desirable for the parts that are to be replaced to have the same life, or a multiple of the life, of those with the shortest life. This is to keep to a minimum the number of times the product must be out of service due to maintenance. For example, if some parts have a life of 1000 hr, other parts should have a life of 2000 or 3000 hr rather than 800 or 1200 hr.

In the interest of proper maintenance at the lowest cost, consideration should be given to the skill that is required of the personnel who are to perform the maintenance. It is best if a minimum of skill is required. Also, an effort should be made to make knowledge of an unusual or intricate mechanism unnecessary. It is best if the maintenance can be performed with standard tools and with equipment that is likely to be available at the facilities where the maintenance will be performed.

5-14 COSTS

It is the responsibility of a designer to provide the required quality at the lowest cost, or if the price of a product is established, to provide the highest quality at that price. It is his further responsibility to so design that the company which employs him can make a profit on the product. Unless both cost and quality are given due attention, the company cannot long survive.

The manufacturer's costs include materials, labor, and a number of other items such as paper work, tooling, handling, and machinery, as well as the many costs of keeping the company going that are little affected by the work of the designer. It may be well to point out that the cost of paper work and handling for a few inexpensive parts may be greater than the price of the parts.

The actual costs for the user consist not only of the price paid for the product but also of the costs of operation, maintenance, and repair. The initial cost is important, but the total cost is of greater importance to the users of products that involve operating and maintenance costs.

Minimum cost, regardless of which type, consistent with other considerations, is something largely achieved in the early stages of design. The attitude, "I will develop something that will work and then think about reducing the cost," is an error, for by the time most of the functions are provided for, many of the best opportunities for reduced cost have been passed up. A great deal can be accomplished in minimizing cost merely by keeping in mind its importance. The basic concept for a low-cost design is, "How can this function be best produced?" not "How can this part be made for less?" A designer must realize that reduced cost does not mean reduced quality. Both unnecessary cost and poor quality are generally the result of poor answers.

When a designer has finished a design, he should analyze it for cost before he announces its completion. He should be able to give genuine answers to many questions that he asks himself, such as: Why use alloy steel here? Must this part be heat-treated? Is corrosion-resistant steel really necessary for this? Could these tolerances be increased? Could the need for special tools be avoided? This part is difficult to remove; should it have a longer life? Does this part have to be painted? Of course, these questions should have been asked before this — and most undoubtedly were. However, it is better to ask all the questions again and to find one for which there is not a good answer, than to have it asked later when parts have been received and tooling has been made.

5-15 BUDGETS AND SCHEDULES

Any design is worth an investment of only so much, and if the cost of developing it is excessive, the company may lose money. In an effort to prevent such a loss, estimates are made and an amount of money that can be allowed for design is determined. This allowance, or budget, often is in the form of so many hours. It is important for a designer to know the time that has been budgeted for a design, so that he can, in turn, budget the time he can spend in the various phases of the design procedure.

A budget should be regarded as a guide or a goal, and an effort must be made to meet it. It is the further responsibility of the designer to inform his supervisor if it appears that he will exceed the budget. A designer should keep an accurate record of the time actually spent on each design. The primary purpose of this record is not to keep track of him, but to serve as a basis for future budgets.

A schedule is concerned with dates. The purpose is to ensure completion of a specified amount of work by a given date. A budget and a schedule are not the same thing, and often there is the question of whether to meet the schedule or the budget. Several people can be assigned to a project in order to complete it by a given date and, in so doing, exceed the budget, whereas if only one were to do it, the budget could be met because of the greater efficiency of one person in producing a design.

A designer must advise his supervisor if it appears that he will not meet the schedule. This must be done as soon as possible, so that action can be taken to prevent the larger schedule from being adversely affected.

Several methods have been developed to facilitate the planning of projects. Methods such as critical-path scheduling and the Program Evaluation and Review Technique (PERT) are used for large, complex projects. A less complicated version can be used to advantage on small projects. The basic idea is: all events are given a number and are placed on a large sheet of paper approximately in the order in which they must be completed, progressing from left to right, as shown in Fig. 5-2. An arrow is drawn from an event to each of those that follow it which may be undertaken when it has been accomplished. The length of the arrows has no meaning. Each arrow represents a job that carries the project from one event to the next. The time required to do each job is estimated and is placed above the arrow. The numbers in Fig. 5-2 represent days.

The critical path may now be determined. It is the direct path from start to finish that requires the greatest total time. This path is shown by the heavy line in Fig. 5-2. The time required by this critical path is the minimum time that is required to complete the project. The critical jobs are those that determine the critical path and are the ones that should be concentrated on in order to meet the schedule.

If something is done to reduce the time required to complete one of the critical jobs, a new critical path may be produced.

5-16 OBSOLESCENCE AND RETIREMENT

In the design of some products, consideration should be given to the situation when the product is to be replaced. A product may be replaced because it has deteriorated to the point that it is not performing properly, or because it has become obsolete, that is, because a new unit can perform better or more efficiently.

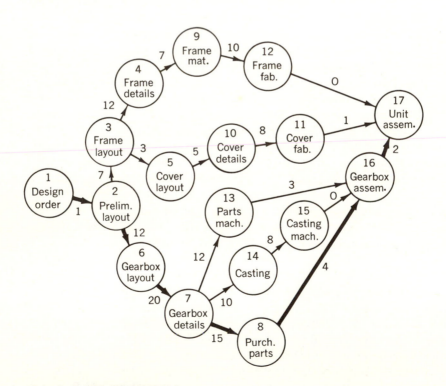

Fig. 5-2 A critical-path schedule.

An important problem faced by the designer is whether to have physical deterioration or technical obsolescence determine the life of the product. The ideal situation, of course, is to have both occur at the same time. However, factors such as accuracy and reliability may require a quality that results in a longer physical life.

It may be appropriate to design so as to facilitate the rebuilding of the product to the original standard at a fraction of the cost of purchasing a new unit. But the product must be one that will not be so obsolete that rebuilding would not be justified. It may be advantageous to design for several levels of use, each less demanding than the previous one in regard to capacity, accuracy, reliability, etc.

Some products should be designed to make salvage more profitable; that is, components with a long life should be easily removed. The recovery of materials may even be appropriate. In some instances, it may be best to so design that the product can be replaced with a minimum of interference with normal operation.

5-17 QUANTITY MANUFACTURED

The number of units to be produced largely determines the manufacturing processes that can be used economically. If a large number of identical parts are to be made, extensive tooling can be used. This tooling will have a high total cost, but when spread over the large number of parts the cost per part is small. A single part may not justify the cost of a special milling cutter, but 100,000 parts could justify the purchase of a special machine.

The number of parts to be made from a particular type and size of material affects the cost of the material on a per-part basis, because of the reduction in price per pound when a large quantity is ordered.

The quantity to be manufactured has considerable effect on the amount of time that can be spent in developing the design, for this cost must be divided among the units produced in the same manner as the cost of tooling. Thus, quantity will have a great effect, not only on the design developed, but on the manner in which the design work is done.

5-18 AVAILABILITY OF PARTS AND MATERIALS

It is the job of the purchasing department to get the parts and materials that are specified. However, it is the responsibility of the designer to consider their availability. The fact that an item is listed in a catalog is no assurance that it is readily available. When time is short and only a few units are required, the less common items are best avoided, even though they would be more desirable than those which will be substituted. It would be much better to design with that which is available than to have to redesign, or to fail to fully satisfy the requirements because of a substitution.

If an item is available from only one source, an investigation of the reliability of that source should be made before it is incorporated in the design. This is particularly important if a great many units are to be produced.

5-19 LEGAL MATTERS

A designer is not expected to have any competence in legal matters. However, he should be aware of several situations where his work may be affected by, or have an effect on, such matters. A designer has not only the responsibility to protect the reputation of his employer, but a moral responsibility to protect those who use the products he designs. There is also a legal responsibility imposed upon a manufacturer which has a great effect on the work of a designer. It is the duty of the manufacturer to so design and make a product that it is reasonably fit for the purpose intended. The manufacturer is expected to have the knowledge and skill of an expert in the field in which he is engaged, and the claim that a hazard was not foreseen will carry little weight. A manufacturer is in an especially poor legal position if he fails to incorporate a safety device that could have prevented injury.

A manufacturer has an obligation to warn the user of any danger involved in the use of his product, even a likely misuse of it. A manufacturer does not like to imply that there is a danger in using his product; thus, a designer should, where it is at all possible, so design that such a warning is unnecessary.

An important bit of evidence in a liability case is that the product conforms to the standards that prevail in the industry. Therefore, any departure from the generally accepted practices that could result in injury should be very carefully considered.

It must be recognized that a drawing used by someone other than employees of the company that a designer is working for may become a legal document, the same as a contract. A drawing in such a situation that can be interpreted in several ways, or that relies on the usual practice in the shop of the designer's company, may cause a great deal of unnecessary trouble and expense.

5-20 PATENTS

This discussion is limited to the situation where the designer is employed by a company that is familiar with patents, rather than the situation of an independent inventor. A patent is the right granted by the United States to a person to exclude others from making, using, or selling that which is patented. In order to obtain a patent, the invention must be new and not merely the application of common knowledge and techniques to a new design situation.

A patent will not be granted if the invention has been used by the public, or if it has been described in a published article more than one year before the application for a patent is made. Only the inventor may make application for a patent — a company cannot, though the inventor may assign his rights to the company. The practice is for the inventor to assign the patent to the company when the invention is the result of work for which he was paid by the company.

There are instances where it is not desirable to patent an invention. The most common is where the demand for the product is expected to be of short duration. Another situation where a patent is of little importance is when the company has decided to put its time and money into producing a superior product rather than into obtaining a patent and defending its rights.

It is best to consider a patent early, if it appears that something worth patenting has been invented. The first step is to make a patent search; that is, try to find a patent that covers that which has just been "invented." It is desirable to begin the search as soon as possible, to avoid developing something that is already patented. The search can also serve as a source of information that will be helpful in the development of the product even if it is not to be patented.

5-21 APPEARANCE

Appearance is a factor, the importance of which varies with the product and the user. If the user is the public and the product is rather common, appearance is very important, and a specialist should be consulted. For less critical situations, a designer can usually produce a design that is satisfactory as far as its appearance is concerned by applying ideas such as those which follow.

A rectangle is more pleasing than a square, and it should appear to be a rectangle, not just prove to be one on measurement. A symmetrical arrangement is more pleasing than an unsymmetrical one, though excessive repetition of items of the same size and shape can produce an undesirable monotony. An object that usually remains in one position should be, and should appear to be, stable in that position. This can be achieved by making the base large enough. It is usually poor to divide the vertical dimension into two equal parts. For example, Fig. 5-3a is more pleasing than Fig. 5-3b, which is equally divided.

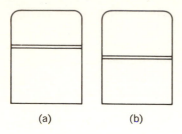

(a)   (b)

Fig. 5-3 The effect of vertical division on appearance.

When working on the appearance of a product, it must be remembered that what appears pleasing in two dimensions may not be acceptable in three dimensions. It would be well to make several pictorial drawings from different viewpoints. These drawings should be based on actual dimensions. An "artist's drawing" will produce an erroneous impression.

A product and its components should appear suitable for the purpose they are to serve. In order to give the impression of adequate strength, it may be necessary to make a part larger than is required by the load it must

Fig. 5-4 A washing machine of modern appearance.
(*Maytag.*)

withstand. An effort should also be made to give the product a modern appearance, which is best described by words such as: smooth, uncluttered, rounded. And the basic shape should be based on a simple form. This is illustrated by the comparison of the washing machine shown in Fig. 5-4 with that shown in Fig. 5-5, which was produced by the same company in 1920.

Another factor in appearance is that which may be thought of as "appropriateness." Two examples of inappropriate practice are: an unfinished sand cast surface on a precision instrument, even though a smoother surface would not better serve the purpose, and a thin sheet metal bracket welded to a heavy cast-iron frame.

5-22 SELECTION OF MATERIAL AND MANUFACTURING PROCESSES

It was stated in Chap. 3 that the designer must select the material, and in Chap. 4 that it is not his responsibility to specify the manufacturing

Fig. 5-5 An old-style washing machine.
(*Maytag.*)

processes. However, the magnitude of the interdependence between the material and the manufacturing processes is such that it is practically impossible to consider them separately. Thus, even if the manufacturing processes were not largely dictated by the configuration of the part, the tolerances, and the quantity to be made, they would be to an extent determined by this interdependence.

It is not the purpose of this section to present a summary or a formula for the selection of the material. The purpose is to present some ideas and concepts that have a bearing on the selection of materials and manu-

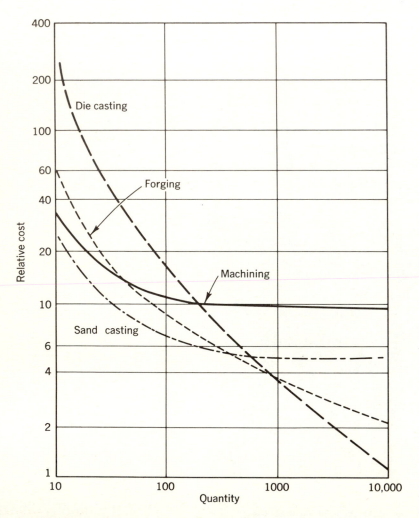

Fig. 5-6 The relation between cost and quantity.

facturing processes that have not been discussed, and to emphasize some that have already been mentioned.

The functional requirements of the part must be satisfied. This involves items such as: part size, configuration, surface roughness, tolerance, ultimate strength, modulus of elasticity, hardness, elongation, thermal conductivity, proportional limit, endurance limit, and abrasion resistance. Production requirements must also be considered. This involves adaptability to processes to be used such as: casting, forging, heat-treating, machining, bending, stamping, and extruding. Also to be considered is the ease with which production can be achieved in the required quantity and in the time that is available.

The designer must fully appreciate the fact that each process has a natural degree of precision that can be maintained, and that to require a greater degree of precision may increase the cost to the point where the process is not economically feasible. Another fact that must be kept in mind is that for each process there is a minimum number of parts that must be produced to make the process economical. This number is not constant for a particular process, for it is affected by factors such as the complexity of the part. However, there is a representative number for the average situation. The relation of cost to the quantity to be made for several methods of producing the same part is shown in Fig. 5-6.

Thought must be given to the adaptability to assembly, which may involve: welding, bolting, brazing, or riveting. Often overlooked is the

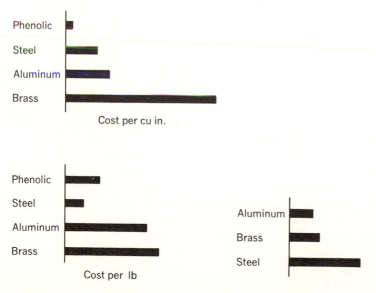

Fig. 5-7 The relative cost of materials.   Fig. 5-8 The relative cost of machining.

adaptability to the desired finish, which may be: plating, painting, polishing, chemical coating, etc.

The cost of most material is based on weight, but it is used on a volume basis. Thus, care must be exercised in making cost comparisons. The relative costs of some common materials, on both the cost-per-pound and the cost-per-cubic-inch basis, are shown in Fig. 5-7. Other relations, such as the cost per psi of strength, or the weight per psi of strength, may be the appropriate criteria for evaluation.

One material is not as readily machined as another. The relative costs of machining several common metals are shown in Fig. 5-8. Comparison of Figs. 5-7 and 5-8 leads to the conclusion that the total cost of a part may be lower if a more costly material is used.

When comparing the costs of materials, the materials must be in the form in which they will be used for the comparison to be a valid one. Also, their availability in the desired form must be considered.

5-23 MAN-MACHINE RELATIONS

A designer must consider a number of factors other than those classed as mechanical or economic. Some of these concern the relation of people to the product. When a product provides information, or has controls that must be operated, it is usually referred to as a machine. In many instances, the consumer or customer to whom the machine is sold does not operate it. In such a case, the consumer and the operator may have different requirements to be satisfied.

In the sections that follow, some basic principles that pertain to the effect the characteristics of an operator have on the design of a machine will be briefly discussed.

5-24 FUNCTION BY MAN OR MACHINE

When a person is required to bring about the successful operation of a machine, he becomes part of a man-machine system. In many instances, man is an essential link in the system. The designer's problem is to determine which of the functions that are required are best performed by the machine and which are best performed by the man. In designing such a system the designer, in a sense, works backward from the objective to be accomplished to the performance of the man. In doing so, he must constantly keep in mind man's capabilities and limitations. It is essential to consider not only what a man can do, but what a man will be willing to do. This will be discussed in Sec. 5-27.

Man is generally superior to a machine in the following important respects:

1.   Man can exercise judgment based on past experience that is not identical to the situation at hand.

2.   Man is very flexible; that is, he can adapt to variations in the manner in which information is presented and respond in an appropriate manner which cannot be completely specified beforehand.

3.   Man can deal with momentary overload, such as being given too much information or too many items demanding control movements.

4.   Man possesses great sensitivity to light and sound.

5.   For the range and complexity of the functions he can perform, man is relatively small, light, and inexpensive.

6.   Man can readily interrupt himself while performing an action, if need be.

A machine is usually superior in the following respects:

1.   A machine can exert very large forces.

2.   A machine can produce motions of extreme accuracy.

3.   A machine is not subject to fatigue or boredom.

4.   A machine responds to signals much more quickly than a man.

5.   A machine can do many things simultaneously.

6.   A machine is not limited by psychological factors; thus, it is not concerned about its safety or its status. Nor does it have the tendency of man to see what he expects to see.

7.   A machine is an excellent monitor of infrequent events.

8.   A machine can perform without difficulty in extreme environments.

In deciding whether the function will be performed by man or machine, consideration must be given to the skill that will be required of the operator. This involves the cost of selecting and training the operator; also to be considered is the availability of people for such training.

A designer cannot change the characteristics of man; thus, he must so design the machine in the man-machine system that the characteristics in which man excels are fully utilized, and so that his limitations are not allowed to cause improper or inefficient performance of the system.

5-25   PRESENTATION OF INFORMATION

The manner in which information is presented has a great effect on the performance of a man-machine system. There are numerous instances where the improper interpretation of information presented to an operator has resulted in disaster. And too many times it has been found that the manner in which the information was presented was responsible for, or contributed to, the improper interpretation.

Only essential information should be presented, and it should be received by the most appropriate sense, which is usually vision. There are several principles governing the arrangement of indicators:

1. The more important indicators are placed in the best location.
2. The indicators that are most frequently referred to are placed in the best location.
3. Similar indicators are placed in the same location.
4. An indicator is so located that the operator does not have to move to another location to perform the control operation required by the information that has been presented.

It is obvious that these principles are incompatible; therefore, the designer must determine which are most appropriate in the situation with which he is dealing. It may be that a combination is most appropriate. In any event, careful consideration of these principles will produce a better arrangement than if no principle of location was followed.

The numerals of an indicator located at a normal distance from the eye should be 0.15 in. high. If the indicator is located more than 2.5 ft from the eye, the size of the numerals and the markings, as well as the size of the pointer, should be increased. The indicator must be adequately illuminated. The amount varies with the spacing of the markings and accuracy required; from 30 to 50 foot-candles usually is adequate. Care must be exercised to prevent glare or reflections. If an indicator is frequently referred to, it should be located in a plane perpendicular to the line of sight when the operator is in his normal position, and it should be at or below eye level. If a number of dials are to be monitored for being in the normal range, they should be arranged so that the pointers are horizontal when indicating the normal condition.

The numerals of an indicator should always be horizontal when in the position to be read. Pointers should not cover the numerals. The scale should increase in a clockwise direction if circular, and increase to the right or up if a straight scale. An indicator of the counter type, such as the mileage indicator of a speedometer, is usually superior to other types, except in instances where the numerals would move so fast as to make reading difficult. Numerals should be about one-half as wide as they are high and should have a stroke width about one-eighth of the height. Letters should be about as wide as they are high and have slightly greater stroke width.

A band of color along the scale can be used to indicate various ranges. Green should be used for the normal range and red for a range that is abnormal or dangerous.

5-26 CONTROL ACTS

The comment concerning the importance of proper design and arrangement of indicators applies to the design and arrangement of controls that the operator must use when performing his function in the man-machine system.

The more important types of controls that are operated by the hand are: knobs, levers, handwheels, cranks, and various types of switches. Those that are operated by the foot are pedals and simple switches. These controls cannot be classified as good or bad, but only as appropriate or inappropriate for the particular function which they are to perform. The problem for the designer is to select the most appropriate, considering a number of factors.

The basic factors to be considered in designing a control are: the force required of the machine to perform the function, the force to be exerted by the operator, the precision required in positioning the control, the number of control settings or the range of a continuous type, and whether the operator must get information from the control.

The principles that governed the arrangement of indicators also apply to the arrangement of controls. They will be repeated here with some additional principles:

1. The more important controls are placed in the most convenient location.
2. The most frequently used controls are placed in the most convenient location.
3. Similar controls are placed in the same location.
4. A control is located so that the operator does not have to move to another location to get the information from an indicator that is required to perform the control operation.
5. Controls that must be operated in sequence are arranged so that movement from one control to the other is smooth, natural, and forms a pattern.
6. The controls are distributed among the four limbs as much as possible, reserving the more critical for the right hand.

These principles are also incompatible, and judgment will again be required in applying them to a particular situation.

It is reasonable to require an operator to put forth the effort to operate an inconveniently located control, but this should not be required for a frequently used or important control. Care must be exercised to eliminate the possibility, when operating a control, of accidentally actuating another control. This is most likely to happen when the control to be operated is inconveniently located.

At this point, it would be well to mention that emergency controls deserve special consideration. They must be easily recognized and accessible, even from rather unusual positions of the operator, and some provision must be made to prevent their accidental actuation. This may be accomplished by incorporating a guard. But it must not interfere with the rapid operation of the control. A guard may also be used to call the attention of the operator to the fact that he is operating an emergency control, especially if only emergency controls are guarded and all are guarded in a similar manner.

A designer must so design the control arrangement that the operator is unlikely to confuse one control with another. This can be accomplished by adequate separation and by making the control handles dissimilar. The designer must also so design the control that it is unlikely to be moved to the wrong position or to a neutral position. This can be accomplished by making the motion of the control compatible with the movement of the indicator associated with the control; also by making the control move in a direction that is regarded by the majority of people as natural or normal for the response that is desired. These natural movements are: a lever (including a toggle switch) moved right means right, on, or increase; a lever moved up means up, on, or increase; a lever moved in the vertical plane, but away from the operator, means on, or increase, and if a vehicle is being controlled, forward; a handwheel or crank turned clockwise means clockwise, on, or increase. The opposite movement, of course, means the opposite response. For foot pedals, there is no natural motion-response relation. The movement required of the operator should be limited to pushing down upon them. The control center shown in Fig. 5-9 incorporates a number of these principles.

Many variables affect the force that an operator should be required to exert in operating a control. Some important factors are as follows:

1.   Size and strength of the operator: It may be that the machine will never have to be operated by a 90-lb woman.

2.   Number of times the force must be exerted: If it is only 3 times per hour it can be much greater than if 3 times per minute.

3.   Position of the operator: An operator can apply a greater vertical force when standing than when sitting.

4.   Manner in which the force is to be applied: A sitting operator can apply a much larger force to a lever at his side when pushing forward than when trying to move it to the left or right. Even the angle between the upper and lower part of the arm makes a difference in the force that can be exerted.

Fig. 5-9   A control center. (*Leeds & Northrup Co.*)

If it appears that the operator will not have the required strength, the mechanical advantage can be increased. If this requires excessive motion along the arc of a lever, or an excessive number of revolutions of a handwheel, an air or hydraulic cylinder or electric motor will have to be provided. If this is done, it is usually easy to provide the control.

In designing to accommodate the operator, it would be well to keep in mind that providing for a person of average size will cause less difficulty than providing for either a large or small person. However, providing for the average in some instances is a poor design. For example: a person is to operate foot pedals while seated. If the location of the pedals and seat height were arranged for the average American man, about half of the men and most of the women in America would experience difficulty or inconvenience. But if the seat height and pedal location were based on a slightly smaller man, a much greater number of people would be accommodated, and a large man would be only slightly affected. Of course, the ideal solution in this case would be to make the seat adjustable.

5-27 OPERATOR ACCEPTANCE

A machine can function perfectly and still be unsuccessful if the operator develops a dislike for it. Some factors to consider in this respect follow: A machine should not unnecessarily endanger an operator who is exercising a reasonable degree of care. This factor can usually be acceptably provided for by the use of guards or covers and provisions to prevent injury in the event of failure of some part of the machine that can normally be expected to fail, such as a belt. A machine should not require speed and accuracy in the movements of the operator to avoid an incident that is unpleasant, such as having a hand struck. An effort should be made to reduce the noise and vibration produced by a machine.

There are also what may be thought of as psychological factors involved in acceptance by the operator. If the operator is dissatisfied with his role in the man-machine system, the success of the system will be impaired. If the operator feels that his skill and ability are not required, or that he is merely a servant to a mechanical monster, he may become dissatisfied. These objections may be overcome by providing the operator with indicators that tell him what is taking place or how the work is progressing, and a control that will allow him to interrupt the process if he thinks it is going out of control. If the operator is provided with an understanding of the operation of the machine, he will be more inclined to feel that he is the master of the machine.

QUESTIONS

5-1 What are some of the important questions that should come to mind when providing for proper function?

5-2 Give several examples where increasing the size of a part could reduce its weight.

5-3 Why must consideration be given to dissimilar metals in contact?

5-4 Discuss briefly several common environmental factors that affect design.

5-5 What are some of the ways in which the noise produced by a product can be reduced?

5-6 What are the two kinds of life that a product has?

5-7 What are the two ways in which life is provided for when a part is subject to conditions such as fatigue or corrosion?

5-8 Explain the fail-safe concept.

5-9 What is interchangeability?

5-10 Why is interchangeability an important consideration in design?

5-11 What are several ways in which interchangeability can be provided for?

5-12 Explain how standardization can reduce the cost of a product.

5-13 Discuss several important items pertaining to assembly that affect the reliability of a product.

5-14 What is the responsibility of a designer with respect to maintenance?

5-15 Explain preventive maintenance.

5-16 What is the basic concept for low-cost design?

5-17 What are some of the questions that should come to mind when a designer analyzes his design for cost?

5-18 What is the difference between a budget and a schedule?

5-19 When designing a new product, why should a designer be concerned with obsolescence?

5-20 Why is the number of units to be made such an important factor in design?

5-21 What is a patent?

5-22 Why should a patent be considered early, when it appears that something has been invented?

5-23 What are several basic concepts that have a bearing on the selection of materials and manufacturing processes?

5-24 How can using a more expensive material reduce the cost of a part?

5-25 What are important points to remember when comparing the cost of material?

5-26 What are some of the ways in which man is superior to a machine?

5-27 What are some of the ways in which a machine is superior to a man?

5-28 List three principles that govern the arrangement of indicators.

5-29 Discuss several principles that are important in the design of indicators.

5-30 Discuss several principles that are important in the design of controls.

5-31 What are some of the important factors in the location of controls?

5-32 What are some of the factors that affect the force that an operator should be required to exert?

5-33 What are some of the physical factors that affect the operator's acceptance of a machine?

5-34 Discuss the psychological factors involved in the operator's acceptance of a machine.

6 DESIGN PROCEDURES

6-1 INTRODUCTION

It is essential that a design be presented in such a manner that a great deal of explanation by the designer is not required. It is also important that the design be completed in as short a time as possible, not only because of the cost of the designer's time, but because of the need for all others concerned with the project to get started. As a designer gains experience, he either develops or becomes acquainted with methods or procedures that facilitate performing the steps of the design process. It is the purpose of this chapter to discuss these procedures that are so essential to the efficient execution of a design. Some of the things discussed will seem superfluous to those with some experience either on the drawing board or in some other phase of the development of a design, but a beginning designer seldom has such experience.

6-2 PREPARATION OF A LAYOUT

A designer can visualize a design only so far; then it becomes necessary to put that part which has been worked out in a form that is more concrete than a mental image, in order to record the decisions and to clarify the remaining problems. This more concrete form is generally a layout. A layout is essentially just an accurate drawing. Sketches are useful, but they cannot record or provide the quantity or quality of information that a layout can. A layout is a very important "tool" in design; it permits the designer to try several solutions quickly and economically. Figure 6-1 is a portion of a

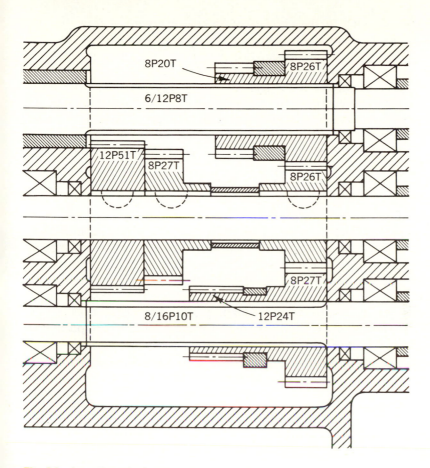

Fig. 6-1 A portion of a layout.

layout which shows part of a gear box by which changes in shaft speed are accomplished by shifting two of the gears.

The purpose of a layout must be kept in mind. It is not an end in itself. It is to be used by the designer in developing his design and in producing the drawings to be used by the shop. There are two attitudes, or tendencies, that must be resisted while making a layout. First, the tendency to seek recognition by producing a drawing that approaches a photograph in the execution of details; or to include views and sections that, though perfect, are unnecessary; or to use unusual practices or techniques. The other undesirable tendency is to regard the layout as a sketch, quickly made to scale, to provide information that the designer needs at the moment.

A beginning designer not guilty of either of these tendencies usually finds it difficult to decide just how much to show and how best to proceed in making his layout. In the absence of specific instructions from the de-signer's employer, the following practices are recommended.

The layout should be made with a relatively hard, sharp pencil on plastic. Consideration should be given to the views that will be required and to their arrangement, also to the scale to use. Usually, the larger the draw-ing, the greater the accuracy. But a drawing can be so large that when using standard methods and equipment the accuracy is reduced. Often, the scale can be increased by removing portions of the assembly where there are no details to be worked out, in which case dimensions must be used to main-tain the relationship of the various features.

After deciding on the scale and arrangement, the basic lines, such as center lines, should be drawn and labeled. This labeling of the lines is im-portant. The designer may think that he will remember what they are, but after many other lines are added there is the possibility of confusion. Also, the designer is not the only one who will work with the layout. If there is any structure or parts that are close to the area that the layout is concerned with, their outline should be shown and labeled. Loads should also be indicated.

After the basic lines and surrounding parts have been drawn, it is usually desirable to place a sheet of cheap tracing paper over the portion of the layout that is to be worked out, and with a soft pencil sketch a solution. Instead of erasing the solution, another sheet of tracing paper is used for the revised solution. This process is continued until the designer is satisfied, at which time the selected solution is added to the layout. This use of tracing paper permits several solutions to be made quickly, and the designer is more likely to consider another design, or a modification, if it does not involve erasing part of the layout that may have to be redrawn exactly as it was because it was the best solution.

Though the layout is only a two-dimensional surface, the designer must continuously think and work on the layout in three dimensions. Something that looks good in two dimensions may be impossible in three. Therefore, consideration should be given to the third dimension as soon as possible, to prevent an unnecessary loss of time. The extreme positions of moving parts should also be shown. Just the outline is necessary, except where details are determined by the positions.

As the layout progresses, the designer should check to make certain that the parts can be assembled. A design may be produced that will work properly, have sufficient strength, consist of parts easy to manufacture, and satisfy all other requirements, except that it cannot be assembled. If the assembly appears to be critical, drawings made on tracing paper of the various parts or tools can be moved about on the layout to simulate as-sembly. If a certain sequence must be followed to accomplish assembly, this sequence should be noted on the layout.

The layout is generally regarded as accurate to ±0.010 in. Thus, any dimension that will have a tolerance equal to or larger than this may be scaled from the layout. All dimensions that cannot be scaled should be given. This, of course, does not apply to standard parts.

All standard parts should be identified on the layout, and the material for any parts to be manufactured should be specified. Also, any heat-treatment or surface finish, such as anodizing or cadmium plating, should be specified. Notes should be used in place of drawing whenever time may be saved, providing clarity is not reduced. When the layout is complete, the designer should put himself in the place of someone who had never seen it before and try to understand it. Some additional sections, views, or notes may be required. Gaining the reputation of producing a good, clear layout is well worth the slight additional effort.

6-3 MODELS

Models are often used in design, because they are an inexpensive means of providing information. Which of several different types to use depends on the kind of information that is desired and the difficulty of obtaining it by some other means.

After the kinematic arrangement of a mechanism has been worked out, and the configuration of each of the elements has been tentatively decided upon, a model of a convenient size can be helpful in checking the design. Errors in the application of kinematic theory are often discovered, as are errors that result from not being able to visualize all related motions simultaneously. Clearances can be checked, or at least the positions where there is a possibility of interference can be easily determined.

For mechanisms whose motions take place in parallel planes, the simplest form of model is a separate sheet of tracing paper for each part, on which has been traced the outline of the part. These sheets are moved about on the drawing board in a manner approximating the motion of the actual parts as nearly as possible. If the mechanism is a simple one, this is all that is required, but the flimsiness of tracing paper parts is a handicap. This handicap can be easily overcome by running prints of the layout, gluing the prints to cardboard, then cutting out the parts. Care should be taken to make the distances between pivot points and the configuration of the parts at the points where there is possible interference as accurate as possible. Thumbtacks are used at the pivot points, and if they are placed with their heads down and a piece of an eraser stuck on them, the parts will remain assembled. If the cardboard parts are not sturdy enough, thin plywood, with bolts at the joints, can be used. Figure 6-2 shows an example, a two-dimensional model of a mechanism to extend an airplane landing flap.

Retracted

Extended

Fig. 6-2   A two-dimensional model.

In the case of mechanisms whose motions are not in parallel planes, a three-dimensional model must be made. Although a small model made from pencils, wire, masking tape, and anything else that may be handy is helpful in visualizing an idea, a sturdy, accurate model must be made if reliable information is required. The parts may still be represented by sticks of wood, rather than by costly carved replicas, but when checking for interference, the configuration of the parts must be reproduced at the points where interference appears possible.

The idea of a scale model can also be used to advantage in the case of small mechanisms, if the model is made larger than actual size. With a carefully made model 10 times actual size, it is possible, in effect, to see 0.001 in.

To determine how well a proposed product fits the person using it, a model of both the product and the person can often be used. The model of

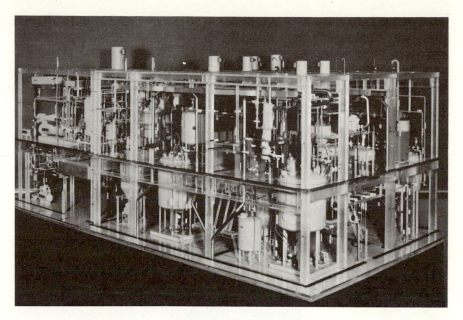

Fig. 6-3 A three-dimensional model. (*Parke, Davis & Co.*)

the person should be constructed with pivots at the joints so that it can be placed in the various positions that may be required. The use of models is the most efficient method of determining the three-dimensional arrangement of components, especially if the arrangement that requires the least volume is desired. Another effective use of models is in presenting a design for approval to people who have limited ability in reading drawings.

The three-dimensional model shown in Fig. 6-3 was used by the contractor, in place of some of the drawings usually prepared, both in preparing the bid and during construction of the chemical processing operation it represents.

6-4 MOCK-UP

A mock-up is similar to a model in many respects, and is sometimes used for some of the same purposes. The major difference is that a mock-up tends to be more realistic, either in appearance or size. A mock-up may be made at one or more stages in the development of the design, depending on the purpose it is to serve. It may be a replica of the proposed product, with the emphasis only on faithfully producing the external appearance. It may be made with the emphasis on simulating instruments and controls, in order to determine the suitability of the arrangement with respect to the

Fig. 6-4 A mock-up. (*Boeing Co.*)

operator. In this case, the appearance may be of no concern. Switches are merely painted in the correct location, and though the controls have the proper location and motion they are roughly represented. Figure 6-4 shows an example, a mock-up of the cockpit of the supersonic transport.

Another type of mock-up is one in which all the components of an assembly are carefully made to their actual size and shape. This type is used to determine whether there are any interferences and to determine that assembly can be accomplished. The various parts of a mechanism may be made so that they can be moved to the required positions, thus checking the kinematic arrangement.

Whenever there is motion in several nonparallel planes, clearances can usually be checked more economically with a mock-up.

A mock-up can be used to advantage in working out the routing of hydraulic tubing or electrical wiring. Tubing and fittings identical to those in the actual assembly should be used. A photograph can be made from the final version, from which a large-size reproduction on plastic can be made. Callouts and notes are added to make it an assembly drawing.

The materials to use in making a mock-up depend on its purpose and size. Cardboard and masking tape are indispensable; clear pine is soft and easily worked, but for small parts where motion is involved, hardwood or soft aluminum would be better. Plastics that soften at a low temperature can often be used to advantage. When actual parts are inexpensive, they should be used.

6-5 PROTOTYPE

A designer is often involved to some degree with the construction of a prototype and the tests performed on it. There are several kinds of prototypes. For one kind, the mechanism will operate but either will not carry the design load or will not have the required life. Another kind has some parts that are not as they will be in the production unit, because of modifications to permit certain tests, or because cost dictates a manufacturing process different than that to be used in production. The most satisfactory form for a prototype is one which is as much like the production unit as possible, and is thus intended to perform the same as the production unit, when subjected to all of the conditions for which it was designed.

A designer may be assigned to supervise the construction of the prototype by marking prints for changes, such as from a die casting to a sand casting, or from a forging to a part that is machined from bar stock. Whenever prints are marked, two of each should be marked identically and the designer should keep one of them. Instructions should be written, and the designer should keep a copy. This does not apply to suggestions or interpretation of the intent of the drawings. He may also be called upon to redesign a part to facilitate construction of the prototype, such as to fabricate a frame by welding which in production will be cast.

In his enthusiasm to construct his design, the designer must not overlook the real reason for the prototype, which is to prove the design of the production units, not merely to make the prototype work. Thus, a change should be made only after careful consideration of all the effects it may have. Any change must be recorded in detail on the layout.

It should constantly be kept in mind that extensive handwork by a skilled craftsman to make the prototype perform properly cannot be allowed in making the production units.

6-6 PROJECT LOG

A designer should make a record of many items pertaining to a design project that do not appear on the layout or in the stress analysis. Items to be included are:

1. The purpose of the project, the date assigned, and by whom it was assigned
2. Specific limitations that apply and suggestions that deserve consideration

3.   References to information that was used in developing the design

4.   Sketches and descriptions of possible solutions and pertinent ideas

5.   Unique advantages and disadvantages, also any special criteria used in evaluation

6.   A description of any major problems encountered in design or development; also the action taken to overcome them

7.   General information such as: specification or drawing numbers assigned, names of persons concerned with the project, and names of subcontractors involved in development

8.   Improvements or areas where investigation may be worthwhile that were not pursued in the present project because of lack of time or a specific limitation

All of these entries should be very brief, except, possibly, item 4. If there is the possibility that an idea which appears in item 4 may be the basis of a patent, it should be fully recorded, and anything at a later date which would indicate that the idea was being developed should also be recorded.

There are several general reasons for keeping such a log: it is an orderly way of going about an important task; it will serve as a source of information that may be difficult to gain in any other way; it is a record of experience and thus a means of avoiding the repetition of an error or an investigation. A specific reason for this record is to serve as evidence in legal proceedings associated with a patent that may result from the design work. In order to best serve the purpose last mentioned, the entries must be made in ink, with no erasures, in a bound notebook which has all the pages numbered. Also, the entries must be dated and periodically witnessed by someone capable of understanding them.

6-7 EVALUATION PROCEDURE

In evaluating several alternative designs or alternative features of a design, a number of items must be considered. These items are not of equal importance, and the designs are not all equally desirable with respect to a particular item. Unless the designs are extremely simple, it is impossible to keep all items, alternatives, and so on, clearly in mind. Thus, a system, or procedure, for evaluation must be used.

The relative merit of a design, or feature of a design, can be indicated by a number, as follows:

NUMBER	MEANING
0	No provision is made in the design.
1	The provision is poor.
2	The provision is regarded as fair.
3	The provision is good.
4	The provision is rated very good.
5	The provision is outstanding.

If these numbers, indicating the relative merit of a design, were added to produce an overall rating, an erroneous conclusion could result. A particular design could have great merit in less important areas which would more than compensate for weaknesses in some important areas. To prevent such an error, the relative importance of an item can be indicated by a number. The number 5 is assigned to an item that is essential and must be provided for by a design or feature rated good or better. The number 4 indicates that an item is important and deserving of considerable effort to provide for in a manner that is rated at least fair. An item that is desirable and worthy of careful consideration is assigned the number 3. Less important items are given the numbers 2 or 1 to indicate their relative importance if the item was listed. If it is known that an item is unimportant, it should not be included in the investigation; however, it may be appropriate to indicate that this item was considered.

If the comparison of the importance of an item and the merit of the design relative to this item is presented graphically, as shown in Fig. 6-5, a clearer and more complete understanding of the situation can be produced. The items that are the basis for the evaluation are placed at the top of a piece of graph paper, and under them a heavy black line is drawn, the length of which indicates their importance. The designs or features that are being evaluated are listed at the left. A light line is drawn to the right, the length of which indicates the relative merit of the design or feature with respect to the item at the top of the sheet. The degree of merit the design has in excess of the importance indicated is shaded in green (colors are indicated by cross-

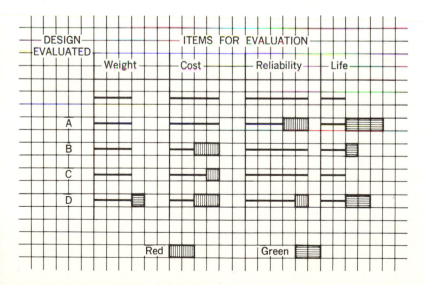

Fig. 6-5 Method of design evaluation.

hatching in the illustration), and the degree the design falls short of the magnitude of importance indicated is represented by shading in red.

Of course, the design with red shading is inferior to one with only green shading. However, additional information is presented:

1. How one design compares with another for any item.

2. For a particular design, the areas that are weak (red shading) and the strong areas (green shading) some of which might be traded to improve the weak features.

3. Any design with more than two units of red under any item is questionable.

4. A design with considerable green shading may be questionable, even though the red shading does not make it so, because the quality of the design is misplaced; thus, it may be unnecessarily costly.

For the simplified situation represented in Fig. 6-5, designs A and D have the highest numerical rating. Nevertheless, design C is found to be the best design in spite of having the lowest numerical rating.

It cannot be emphasized too strongly that a system such as that just described is not precise. Its primary value is that it tends to force careful consideration of the items, because all items must be rated. It also records the decision, and presents a complete "picture" of the evaluation.

QUESTIONS

6-1 What are the most important functions of a layout?

6-2 What are the two extremes to be guarded against in preparing a layout?

6-3 What dimensions are placed on a layout?

6-4 What is the simplest form of model?

6-5 Is there any reason to make a model larger than actual size?

6-6 What are four uses for a model?

6-7 What are four uses for a mock-up?

6-8 What is the difference between a model and a mock-up?

6-9 How does a prototype differ from a mock-up?

6-10 What are the two most important points to keep in mind when changing a prototype?

6-11 What are some of the reasons for keeping a design log?

6-12 List five items that are appropriate to record in a design log.

6-13 What are the reasons for using a rather extensive procedure for evaluating alternative designs?

7 STRESS ANALYSIS

7-1 INTRODUCTION

One of the important responsibilities of a designer is to provide adequate strength in his design. The purpose of this chapter is to discuss the general principles involved in the analysis of stress as *applied* to actual design problems; that is, to correlate the information gained in earlier study with practical considerations. The determination of strength is part of each of the chapters devoted to machine elements. This discussion will serve as the basis for investigation of strength in these particular situations.

In the study of strength of materials the value sought was the load or the stress. A designer is interested in such values, but his primary concern is the determination of the size of a member to withstand the load imposed and not exceed a specific stress. This chapter is intended to aid in this necessary transition of thought.

A member, of course, must not fracture or be permanently deformed, but it also must not deform excessively at any time when subjected to the loads developed in normal operation. Thus, the maximum load must be determined and used as the basis of strength calculations, not an average. An average is often used in other design calculations, but if an average were used in determining strength, the part might be greatly overstressed when the maximum load occurred.

Loads are external forces that are applied to a member and produce stress which is internal. Loads may be classified according to the stress produced, such as tension or shear, and also according to the manner of application, such as steady, fluctuating, or reversing. In this text, the loads are always in pounds, the areas involved are in square inches. Thus, the stress is in pounds per square inch.

A beginning designer often has difficulty in recognizing the stresses involved in an actual situation. The analysis of a situation to determine the stresses that are present is best illustrated with examples, and in the following chapters this will be done when actual design situations can be used.

7-2 TYPES OF LOADS

The type of load is as important as the magnitude of the load in determining the size a part must be to carry the load. A static load is one that does not vary in any way. Because static is associated with the idea of no motion being involved, the term steady load is sometimes used, but the meaning is the same. Also, loads that vary infrequently are treated as static loads. A varying load is one the magnitude of which changes; for example, from a low-tensile stress to a high-tensile stress. A reversing load is one in which the magnitude not only varies from maximum to zero but reverses, such as from maximum tension through zero to maximum compression and back to tension. Shock loads are those that involve impact and that induce stresses higher than if the load were applied gradually. For example, if a part were to stop a falling weight, it would be subjected to a shock load, the magnitude of which would increase with the distance the weight dropped as well as with the size of the weight. If the weight fell only a small fraction of an inch, the height and velocity involved would be negligible, but the load would be considered as suddenly applied. The stress induced would be momentarily twice as great as it would be a few seconds later. Severe shock is seldom found in mechanical design, because it is undesirable for reasons other than strength, such as vibration and noise.

For those loads that are not static, the number of applications or reversals of the load in the expected life of the part becomes important. If this number is large, fatigue is a factor, and strength must be based on the fatigue strength or endurance limit of the material.

7-3 TYPES OF FAILURE

When a part made of a brittle metal, such as cast iron, is subjected to a static load that causes a stress in excess of the strength of the metal, it will deform slightly and suddenly fracture. When a part made of a ductile metal, such as steel, is subjected to such a load, it will deform to such an extent that the load will be relieved by the changed configuration of the part, or destruction of the mechanism will take place, because the deformation allows the parts to interfere with each other. When a part no longer fulfills the requirements imposed upon it by the design, it has failed, even though it has not fractured. A ductile metal under certain conditions will fail in the manner typical of a brittle metal; one of the most common of such conditions is that which leads to fatigue failure.

It is evident from the above discussion that the details of the investigation of strength depend not only on the load but on whether the metal is ductile or brittle. A brittle metal is usually defined as one that has an elongation less than 5 percent in 2 in. It should also be remembered that for a brittle metal the ultimate compression strength is often greater than the ultimate tensile strength, rather than equal to it as is assumed for ductile metals.

If a part is considered to have failed when it does not perform properly under the load that is applied, it may fail when the stress is relatively small if the deflection is excessive. It must be remembered that deflection depends on the modulus of elasticity of the metal, not on its strength. Therefore, a condition of excessive deflection of an aluminum part cannot be corrected by using a stronger aluminum, but it can be corrected by using a metal with a greater modulus of elasticity, such as steel.

7-4 FATIGUE AND ENDURANCE

Parts designed on the basis of ultimate strength or yield strength are satisfactory for static loads, but will likely prove unacceptable in situations where they are subject to many stress cycles. If the number of stress cycles is known, the fatigue strength for that number of cycles may be used. The intention in the design of many parts is that failure will not be due to fatigue. In such cases, strength determinations must be based on the endurance limit.

The endurance limit is found by subjecting polished specimens to reversed bending. For wrought steel, it is equal to about one-half the ultimate strength. If the surfaces are not polished, the value for the endurance limit will be smaller. This will be referred to as the corrected endurance limit. The rougher the surface, the greater will be the difference between the endurance limit and the corrected endurance limit. Also, a given roughness will have a greater effect if the steel is heat-treated.

For parts which are made of low-carbon steel and have the usual finish machined surface, the corrected endurance limit may be as high as 90 percent of the endurance limit. For heat-treated steel parts with a finish machined surface, the corrected value will be from 70 to 75 percent of the endurance limit. For very rough surfaces, the percentage may be less than 25 percent. Corrosion produces a rough surface and a very great loss of strength, and it should, therefore, be prevented, especially when the parts are heat-treated.

For cast iron and cast steel, the endurance limit is approximately 40 percent of the ultimate in tension. For nonferrous metals, it is often assumed to be between one-fourth and one-third of the ultimate.

For reversed axial loads, the endurance limit is about 70 percent of the value for reversed bending, and for reversed torsion or shear, it is about one-half. If the loads are not completely reversed, the endurance limit is higher.

The low percentage given for rough surfaces is not intended for use in determining the size of a part, but to show the importance of surface roughness where fatigue is a factor. Fatigue failures occur most often in parts subject to bending or torsion, but seldom when the loading is axial, especially compression. Brittle metals are usually regarded as unsatisfactory where fatigue is a factor. When cast iron is subject to a stress that is of varying magnitude but is always compression, the endurance limit will be several times as high as that given above.

7-5 STRESS CONCENTRATIONS

Strength in situations involving a great many stress cycles is largely dependent on whether stress concentrations are present, and on their severity. A stress concentration is a situation in which the actual stress is higher than that calculated with the basic equations used in strength of materials — for example, stress is equal to the load divided by the area.

Stress concentrations have many causes, such as: flaws in the material, accidental nicks or scratches, tool marks, and abrupt changes of section. Abrupt changes of section can be very serious, and it is entirely under the control of the designer.

A stress concentration may be visualized by thinking of lines that are straight and evenly spaced where the stress is equally distributed, as shown in Fig. 7-1a at section A-A. In the presence of a stress concentration, the lines are not straight and are closer together, as shown in Fig. 7-1a at section B-B. The greater the departure from a straight line, and the closer together the lines become, the greater the stress at that point. This is evident from comparison of the lines in Fig. 7-1a, c, and e with those in Fig. 7-1b, d, and f, which indicate the magnitude of the stress to scale at sections B-B, C-C, and D-D.

The static load in a brittle metal that has a stress concentration will produce the full increase in stress. If this exceeds the strength of the metal, a crack will develop, which then becomes a more severe stress concentration. The crack also reduces the area carrying the load. Thus, fracture of the part takes place rapidly and without warning. A static load in a ductile metal that has a stress concentration will cause the metal to yield where the stress exceeds the yield strength of the metal. This yielding causes a slight modification in the configuration of the part, and a slight redistribution of the

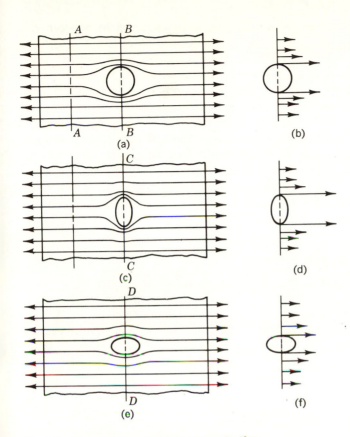

Fig. 7-1   Visualization of stress concentration.

stress. Thus, the local stress is reduced to a value below the yield strength
of the metal.

A ductile metal subjected to repeated loading behaves differently at a
stress concentration, because the fatigue strength or endurance limit is involved. If the maximum stress at a stress concentration exceeds the fatigue
strength or endurance limit, a very small crack is likely to develop, which in
time will increase to the point where there is not sufficient area to carry
the load.

In making allowance for a stress concentration, a number known as a
stress-concentration factor is used. It may be defined as the actual stress
that exists as a result of the presence of the stress concentration divided by
the stress that would exist without the stress concentration. The symbol
used is K. These factors are to be found in books dealing with fatigue.
Where care is exercised in reducing the stress concentrations caused by the

configuration of the part, the factor in most instances will be between 1.2 and 2.0.

Consideration of the comment on visualizing a stress concentration in connection with Fig. 7-1 indicates that an effective method of reducing their effect is to make the transitions in shape and size of cross sections gradual. This is usually accomplished by adding material to the configuration that would be desirable if stress concentration were not a factor. For example, Fig. 7-2a is what is desired, but Fig. 7-2b is used. It may also be accomplished by removing material; for example, Fig. 7-2c, where the groove causes a stress situation similar to that where a large fillet radius is used.

Detailed methods of reducing stress concentrations and appropriate factors will be presented in the chapters on the various machine elements.

7-6 WORKING LOAD

In order to determine the size of an element, regardless of whether it is to be designed or selected, the maximum load that it is expected to carry must be known. The maximum stress that is to be permitted must also be known if a part is to be designed. The term factor of safety is sometimes used in connection with the determination of the size of a part, but because of the confusion that exists relative to this term, it will not be used in this text.

The term working load will be used for the maximum load that the part is expected to carry; that is, the maximum load that will occur in normal operation. The design load will be the working load times any factors that may be appropriately applied to produce the load upon which

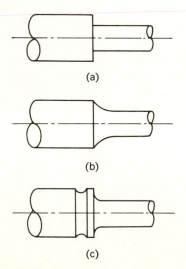

(a)

(b)

(c)

Fig. 7-2 Methods of reducing stress concentration.

the selection of an element is to be based. The design of some parts will be based on this design load, rather than on the design stress; in which case, allowance will have to be made for some of the strength factors yet to be discussed. In other words, the working load is the actual maximum load to which no factors have been applied, and the design load is a working load that has been increased by applying the appropriate factors.

7-7 DESIGN STRESS

The term design stress refers to the stress that is used in determining the size of the part. Thus, the area of a rod in tension will be equal to the working load divided by the design stress. The design stress will be found by dividing the appropriate strength of the material by the design factor.

The strength of the material that will be appropriate depends on the kind of material and the type of loading. For a brittle metal subject to a static load, it will be the ultimate strength, and for a ductile metal subject to a static load, it will be the yield strength. If repeated loading is involved, the fatigue strength or the endurance limit will be appropriate. And if temperature is involved, it will be the strength at the required temperature.

The design factor will be the product of several applicable strength factors. These strength factors will be designated by letters and are defined as follows:

The factor a concerns the material and requires consideration of:

1. The kind of material. For brittle metals, this factor should be at least 2.
2. The quality of the material.
3. Allowance for residual stress that may result from the manufacturing process.
4. The effect of time and environment on the properties of the material.

For a good-quality low-carbon steel bar that is cold-finished and is not subjected to a severe environment, this factor may be 1.

The factor b concerns the load and requires consideration of:

1. The manner in which the load is applied. If it is applied gradually, the factor is 1; if suddenly applied, the factor is 2. Often, the load is applied in a manner such that a factor between 1 and 2 is appropriate. If considerable shock is involved, the factor should be 3 or more, depending on the severity of the shock. (Repeated loading is provided for in using the appropriate strength of the material.)
2. Allowance for a considerable overload.
3. The loads produced by manufacturing processes.

The factor c is associated with the stress concentration K and the surface roughness. The stress concentration factor must be used for brittle metals, even when subject to a static load, and for all metals when repeated loading is involved. An allowance can be made for surface roughness by reducing the appropriate strength of the metal, as discussed in Sec. 7-4.

The factor d allows for the uncertainties that may exist. The following items should be evaluated carefully, keeping in mind that a factor of 2 reduces the design stress to one-half and increases the size appreciably.

1.   If failure would endanger life, cause a considerable loss of property, or cause a costly loss of time, a larger factor should be allowed for uncertainties.

2.   A high-quality product requires a larger factor for any unavoidable uncertainties.

3.   An important question is the degree of uncertainty involved in the mathematical analysis, or in the determination of the working load, or in the strength of the material. In situations of better than average certainty, a factor of 1.2 is allowed; for average situations, 1.6. For situations of less than average certainty, a larger factor should be allowed, the amount depending upon the degree of uncertainty.

4.   Consideration of some factors discussed in Chap. 5 may also be appropriate in special cases. It is essential to allow for all factors that are applicable. However, care must be exercised so that allowance is made for each factor only once.

The method of determining the required size when designing a part may be summarized as follows:

$$\text{Design factor} = a \times b \times c \times d \tag{7-1}$$

$$\text{Design stress} = \frac{\text{appropriate strength of material}}{\text{design factor}} \tag{7-2}$$

$$\text{Size of part} = \frac{\text{working load}}{\text{design stress}} \tag{7-3}$$

For the alternate method in which the design load is used,

$$\text{Size of part} = \frac{\text{design load}}{\text{appropriate strength}} \tag{7-4}$$

It should be mentioned that the strength of some mechanical elements is dictated by codes, the equations and strength values for materials being specified. Great care must be exercised in using these equations with strengths other than those specified, because a factor may have been included in arriving at the strength. The same caution also is necessary in using the strengths given in equations other than those specified.

7-8 COMBINED LOADING

Many mechanical parts are subject to loads that simultaneously produce more than one type of stress. These stresses must be combined in order to determine the size required — for example, in a shaft subject to bending and torsion. Stresses that are similar to the extent that they act in the same direction can be combined by applying the principle of superposition; that is, they can be added algebraically. Tensile stress is regarded as positive, and compressive stress as negative. Thus, a simple beam to which a tension load is also applied has a maximum tensile stress equal to the tensile stress due to the tension load plus the tensile stress due to the bending moment. The maximum compressive stress will be the maximum compressive stress due to the bending moment minus the tensile stress due to the tension load.

When the stresses are such that the principle of superposition cannot be applied, as is the case with combined tension and shear, the situation involves a theory of failure. These theories are discussed in detail in texts on the strength of materials. The two most commonly used are the maximum-normal-stress theory and the maximum-shear-stress theory.

The maximum-normal-stress theory states, in effect, that failure occurs when the maximum normal stress exceeds the tensile yield strength. The equation for this stress is

$$s_{t,\max} = \frac{s_t}{2} + \sqrt{s_s^2 + \frac{s_t^2}{4}} \tag{7-5}$$

where $s_{t,\max}$ = maximum normal stress, psi
  s_t = tensile stress, psi
  s_s = shear stress, psi

The maximum-shear-stress theory states, in effect, that failure takes place when the maximum shear stress reaches the yield strength in shear. The equation for this stress is

$$s_{s,\max} = \sqrt{s_s^2 + \frac{s_t^2}{4}} \tag{7-6}$$

where $s_{s,\max}$ = maximum shear stress, psi.

Again, the equation to use depends on whether the metal is brittle or ductile. For brittle metals, the maximum-normal-stress theory is the most accurate, and for ductile metals, the maximum-shear-stress theory is recommended.

7-9 PROCEDURE FOR STRESS ANALYSIS

The calculations, assumptions, and decisions that determine the size of a mechanical element should be presented in a manner that will inspire confidence in the person who made them. It should be easy for someone other than the designer who prepared them to follow what was done. Thus, each page should be numbered and should state exactly what it pertains to. Reference should be made to all important information used in making a decision or assumption. If data or equations are not those commonly used, their source should be given. Sketches should be used to define symbols and to indicate dimensions. Equations should be given in symbol form, and the meanings of the symbols with the required units should be specified.

Another reason for doing the stress analysis in a methodical manner is to reduce the possibility of making a required allowance more than once or, what is worse, neglecting to make an essential allowance.

The beginning designer must realize that in designing a part the proper size cannot be arrived at by applying a few rules. Recommendations are helpful, but ultimately, the determination of the appropriate degree of strength is dependent on the judgment and experience of the designer.

QUESTIONS

7-1 What is the difference between a varying load and a reversing load?

7-2 What is the difference between a suddenly applied load and a shock load?

7-3 Compare the mode of failure for a part made of a brittle metal with the mode of failure for a part made of a ductile metal when the failure is caused by a tensile stress.

7-4 Under what conditions will a ductile metal fail in a manner similar to a brittle metal?

7-5 Explain how a part can be considered to have failed even though it did not fracture.

7-6 Discuss the effect of heat-treating and surface roughness on the endurance strength of steel.

7-7 What is a stress concentration?

7-8 What are some of the causes of stress concentrations?

7-9 How does a brittle metal act at a stress concentration when the load is static?

7-10 Describe the behavior of a ductile metal at a stress concentration when the load is static.

7-11 Discuss the behavior of a ductile metal at a stress concentration when subjected to repeated loading.

7-12 Define the term stress concentration factor.

7-13 What are two factors pertaining to material that should be considered in determining the design stress?

7-14 What are two factors pertaining to the load that should be considered in determining the design stress?

7-15 What are two factors pertaining to uncertainties that should be considered in determining the design stress?

7-16 What are two reasons for preparing the stress analysis in a methodical manner?

7-17 What are some of the details that are important in the preparation of the stress analysis?

SELECTED REFERENCE

Roark, R. J.: "Formulas for Stress and Strain," McGraw-Hill Book Company, New York, 1954.

8

DIMENSIONS, TOLERANCES, AND FITS

8-1 INTRODUCTION

The proper method of placing dimensions on a drawing is covered in drafting texts and will not be treated here. What will be considered is how the size, shape, and location of the features of a part are to be controlled. A designer is not only concerned with determining the critical dimensions to assure proper assembly and function, but with providing for tolerances that are great enough to permit economical manufacture.

Because of the increasing importance of surface roughness, it will be discussed in considerable detail.

8-2 DIMENSIONS

The surfaces of a part may be classified as functional or nonfunctional. Nonfunctional surfaces are those which usually do not come in contact with other surfaces in the assembly or operation of the machine, or if they do, their location is of little importance. Examples are the rough surfaces of castings or forgings, and bolt heads.

Functional surfaces may be either primary or secondary. Primary surfaces are those that control the position or motion of parts. Holes are often used as primary surfaces which have the characteristic of locating in two directions. Secondary surfaces are those that, though they do not control position in the sense of a primary surface, must have a particular characteristic that makes a nonfunctional surface unacceptable. An example would be a spotface.

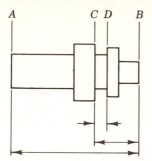

Fig. 8-1 Relation of dimensions.

Surfaces may have a direct or first-degree relationship where there is only one dimension between them, as *A* and *B* in Fig. 8-1. The relation of indirect dimensions may be second degree, as *A* and *C* in Fig. 8-1, or third degree, as *A* and *D* in Fig. 8-1.

A certain relationship between surfaces is required when they are primary or secondary functional surfaces. There is also a relationship that exists between surfaces which is dependent upon the manufacturing processes involved. For example, surfaces tend to be more accurately related when they are produced by operations performed simultaneously or during the time a part is securely clamped in a machine or fixture. Therefore, the designer should try to utilize the relationship between surfaces that results from manufacturing operations for primary functional surfaces that require an accurate relationship.

Whenever a close relationship must be maintained between two surfaces, they should be directly related. For if they have an indirect relationship, several dimensions will have to be held much more accurately than would otherwise be necessary. For example, in Fig. 8-2 dimension *C* should be used, rather than *A* and *B*, if the width of the slot had to be closely held.

Another idea that should be kept in mind is that if the width of the slot *C* in Fig. 8-2 is critical and must be accurate, this does not require that *A*, *B*, or *D* must also be closely held.

8-3 TOLERANCES

In order to produce parts in large quantities and thus gain the advantage of lower cost, they must be interchangeable. This requires some method of controlling the size. It is not economically feasible to make parts to exact dimensions, and in most instances it is necessary to have only a few dimensions closely held. The general concept of tolerance is that it is the acceptable magnitude of variation from the exact size. Because a designer is responsible for providing for the variation that must be permitted in the manufacture of parts, he must be the one to specify the magnitude of variation that is acceptable.

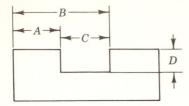

Fig. 8-2 Effect of method of dimensioning on tolerance required.

Because of the great effect that tolerance has on cost, it should be as large as possible. This requires additional effort on the part of the designer to make certain that it will not cause difficulty in assembly or operation. It is much more desirable for the designer to put forth this effort when developing the design than for him to take the attitude, "I will hold the tolerance close, and if it causes trouble in manufacture I will try to increase it."

If the part shown in Fig. 8-3a is a ½-in. rod, the 0.500 is referred to as a nominal size. If the 0.500 is the size that would be specified if tolerance were not necessary, it is referred to as the design size. The tolerance that is permitted may be specified in several ways. The method shown in Fig. 8-3b is the unilateral system (the tolerance is all in one direction). That shown in Fig. 8-3c is the bilateral system (the tolerance is allowed in both directions). In Fig. 8-3d the tolerance is indicated by specifying the limits, that is, the maximum and minimum sizes that are acceptable. The precise definition of the term tolerance is: the *total* variation that is permitted. Thus, in Fig. 8-3 the tolerance in each case is 0.010.

Though these are just different methods of saying the same thing, it should be noted that the method shown in Fig. 8-3c may be used to indicate the desired size. This method is appropriate only in special situations.

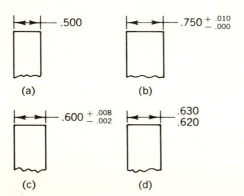

Fig. 8-3 Tolerance systems.

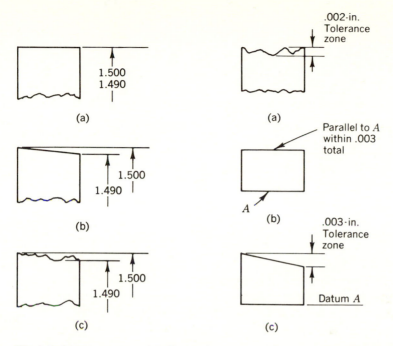

Fig. 8-4 Interpretation of a dimension.   Fig. 8-5 Geometric tolerance.

8-4 GEOMETRIC TOLERANCES

The interpretation of the dimension shown in Fig. 8-4*a* allows the conditions shown in Fig. 8-4*b* and *c*. The design may be such that the variation in size is acceptable, but not the lack of parallelism shown in Fig. 8-4*b* or the lack of flatness shown in Fig. 8-4*c*. To avoid extremely close tolerance merely to control the form of a part, geometric tolerances may be used. In this system a note may specify that a surface is to be "flat within 0.002 total." The interpretation is shown in Fig. 8-5*a*; the entire surface must lie between two parallel planes 0.002 in. apart. To avoid the condition shown in Fig. 8-4*b*, a tolerance on parallelism as shown in Fig. 8-5*b* would be used, the interpretation of which is shown in Fig. 8-5*c*. The surface must lie between two parallel planes 0.003 in. apart, which are parallel to the datum plane *A*. A datum plane is defined as shown in Fig. 8-6. In Fig. 8-6*a* the datum plane is established by the high points of the surface. If there is a considerable tendency for the part to rock, the datum plane is established as shown in Fig. 8-6*b*, where the two distances marked *X* are equal.

A geometric tolerance applies, regardless of feature size, unless stated

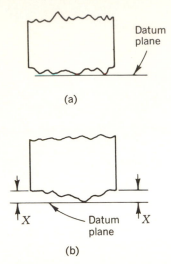

(a)

(b)   **Fig. 8-6 Interpretation of datum plane.**

otherwise. Geometric tolerances should be specified only where the size tolerances permit a condition that will be detrimental to the success of the design.

8-5 POSITIONAL TOLERANCES

When the commonly used system of dimensions and tolerances shown in Fig. 8-7a is used to locate several holes, the condition shown in Fig. 8-7b is possible (the tolerance is exaggerated). The two extreme positions of the upper-left-hand hole are shown by the two dotted circles. It is apparent from consideration of this figure that to make allowance for this possibility in determining the size of the holes required will be rather involved, and that the hole will have to be oversize by a considerable amount if reasonable tolerances are to be permitted.

The design may permit considerable tolerance in the initial location of a part to be attached by the use of bolts in holes — four holes in the case shown — but once located, it must be limited to only a small variation from that position. In such a situation, the method shown is obviously undesirable.

A method of avoiding such situations is to use true-position dimensioning. In this system, the dimensions locating the holes have no tolerance, but the holes are then permitted to vary from this exact location by a specified amount. This is shown in Fig. 8-8. The shaded circles indicate a tolerance zone of 0.010 diameter, within which the center of the hole must lie. The two extreme positions of the upper-left-hand hole are shown by dotted circles.

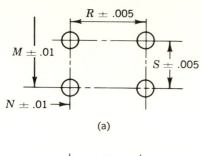

(a)

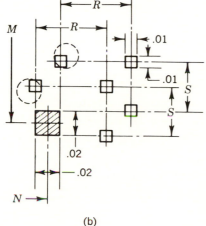

(b)

Fig. 8-7 Interpretation of location dimensions.

Because there must be a tolerance on the diameter of the holes, the question arises whether the tolerance on the holes can affect the location of the holes. If the true position applies when the holes are smallest, assembly would be possible with larger holes when the distance between them exceeded the tolerance allowed. Generally, this additional tolerance is acceptable and the true position is considered to apply to the maximum material condition (MMC). But if this additional tolerance is not to be permitted, the notation RFS is used, which means that the true position is to be held regardless of feature size.

There is another method of dealing with the problem of locating holes in parts that must assemble, with which a designer should be familiar. In this method, a note is placed on each set of holes stating that they must match the other set. This shifts to manufacturing the responsibility for the parts being made so as to ensure proper assembly. But it may also permit achieving the desired result with less effort on the part of manufacturing, for the critical portions of the tooling for both parts can be made simultaneously. This assures proper matching without having to work to very close tolerances.

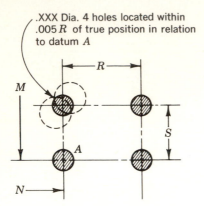

.XXX Dia. 4 holes located within
.005 R of true position in relation
to datum A

Fig. 8-8 True-position dimensioning.

8-6 TOLERANCE STACKS

In the final stages of a design, a layout is carefully drawn to scale using the mean dimensions and locations. The layout shows a clearance where required, and if the parts were made to the mean dimension, the actual

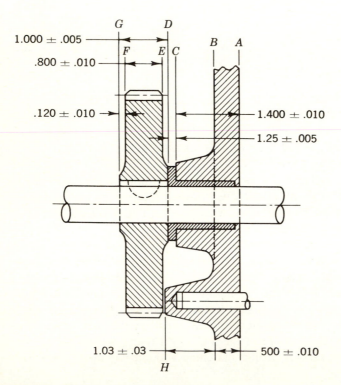

Fig. 8-9 A tolerance stack situation.

clearance would be close to that shown. However, the tolerances that must be allowed may combine to create an interference. Tolerance can also combine to create an excessive clearance. Such unsatisfactory conditions are referred to as tolerance stacks. Calculations should be made wherever an interference or excessive clearance appears possible, to determine if a tolerance stack exists.

A procedure that will facilitate such a calculation will be explained, using the situation shown in Fig. 8-9.

1.   Upon checking this portion of a layout, it appears that there could be an interference between the gear and the boss.
2.   The dimensions that are involved are changed, where necessary, to the form with a balanced bilateral tolerance. The source of these dimensions is the detail drawings of the various parts.
3.   The dimension desired is that between surfaces H and E. To get this, it will be necessary to trace the combination of dimensions that determine it: H to B, B to A, A to C, C to D, D to G, G to F, and F to E.
4.   In the above tracing, if the apparent movement is to the right, the dimension is considered positive, and if to the left, negative. Thus, H-B and B-A are positive; A-C and C-D are negative.
5.   A table such as Table 8-1 is prepared. In the first column, the dimension is identified by part description or letters as used here.

TABLE 8-1   Tolerance Stack

IDENTI-FICATION	DIMENSION		TOLERANCE ±
	+	−	
H-B	1.030		0.030
B-A	0.500		0.010
A-C		1.400	0.010
C-D		0.125	0.005
D-G		1.000	0.005
G-F	0.120		0.010
F-E	0.800		0.010
	+2.450	−2.525	±0.080

MEAN CLEARANCE	INTER-FERENCE	MAX. CLEARANCE
−2.525	−0.075	−0.085
+2.450	+0.080	−0.080
−0.075	+0.005	−0.155

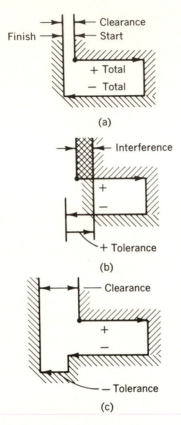

Fig. 8-10 Effect of tolerance accumulation.

6. The dimension *H-B* is to the right. Therefore, it is placed in the + column. The tolerance is placed in the tolerance column. *B-A* is to the right. *A-C* is to the left; therefore, it is placed in the − column, and its tolerance in the tolerance column. This continues to *F-E*.

7. The figures in the three columns are added.

8. The negative total is larger, indicating a clearance. The amount is equal to the difference between the negative total and the positive total, or 0.075. This is the mean clearance. Figure 8-10*a* will help clarify this.

9. The total tolerance is applied to the mean clearance. The + tolerance will decrease the clearance; see Fig. 8-10*b*. And the − tolerance will increase the clearance; see Fig. 8-10*c*.

10. The maximum clearance will be 0.155, and the "minimum clearance" will be an interference of 0.005. Thus, there is a tolerance stack between the gear and the boss.

This unsatisfactory condition can be corrected in several ways. The most obvious and easiest would be to reduce the height of the boss. Assume that this is not possible, that the gear must remain in the position shown, and that the tolerances are not to be reduced. There is probably no reason for the gear being dimensioned in the manner shown; that is, another method of dimensioning that would produce essentially the same part would be acceptable. The mean distance between surfaces D and E is $1.000 - (0.800 + 0.120) = 0.080$. If this dimension were used rather than the 0.120 between F and G, and the tolerance was the same as for the 0.120 dimension, the calculations for possible tolerance stack would be as shown in Table 8-2. The minimum clearance is 0.010. There is no tolerance stack. This clearly shows the importance of carefully considering the manner in which a part is dimensioned.

8-7 FITS

There are instances in mechanical design where the parts, when assembled, must have a certain amount of looseness, or play, between them. In other instances, it is required that the parts be so made that force must be used to assemble them. When looseness is present, it is known as a clearance fit, and the difference between the two diameters is the clearance.

TABLE 8-2 Tolerance Stack

IDENTI-FICATION	DIMENSION		TOLERANCE ±
	+	−	
H-B	1.030		0.030
B-A	0.500		0.010
A-C		1.400	0.010
C-D		0.125	0.005
D-E		0.080	0.010
	+1.530	−1.605	±0.065

MEAN CLEARANCE	MIN. CLEARANCE	MAX. CLEARANCE
−1.605	−0.075	−0.075
+1.530	+0.065	−0.065
−0.075	−0.010	−0.140

When force must be used, it is an interference fit, and the difference between the diameters is the interference.

Because of the tolerance on parts that are to be interchangeable, there will be a variation in the amount of clearance or interference. The maximum material condition is when both the shaft and the hole have the maximum amount of material; that is, when the shaft is at its largest size and the hole is at its smallest size. The minimum material condition is the reverse; that is, when the shaft is smallest and the hole is largest. For either a clearance fit or an interference fit, the tightest fit will be the maximum material condition, and the loosest fit will be the minimum material condition. The difference between the two diameters at the maximum material condition is technically known as the allowance.

Example 8-1 A shaft is to fit in a hole reamed to 1.4995/1.5005. The minimum clearance is to be 0.0005, and the tolerance on the shaft is ±0.001. What is the size of the shaft and the maximum clearance?

Solution The minimum clearance will occur when the hole is its smallest dimension and the shaft is its largest dimension. The largest shaft dimension is, therefore, the smallest hole dimension minus the minimum clearance; that is, 1.4995 − 0.0005 = 1.4990. The smallest shaft will be 0.002 less than the largest. Thus, the shaft size is 1.499/1.497. The maximum clearance will occur with the smallest shaft and the largest hole. Thus, 1.5005 − 1.4970 = 0.0035.

In order to reduce costs, when determining the sizes involved in a fit, consideration must be given to standard sizes. Most often, the design requires machining of the shaft for various reasons, such as providing shoulders for location, and so on. In this case, the size of the hole should be that which can be produced with standard tools and checked with standard gages, and the size of the shaft should be calculated to produce the desired allowance. However, if a considerable quantity of shafting is to be used that does not require machining for reasons other than to provide the proper fit, it is appropriate to use standard shafting and to calculate the size of the hole to provide the required allowance.

8-8 DESIGN OF CLEARANCE FITS

The subject of fits as applied to plain bearings is more appropriately discussed in Chap. 12. Thus, only those fits where assembly, alignment, or position are the determining factors will be covered here. Often, too little consideration is given to these fits, because of their apparent simplicity.

It must be kept in mind that when a clearance fit is used the center of the hole and the center of the shaft will never coincide when in operation. When a setscrew is used, the condition shown in Fig. 8-11 is produced.

The amount of clearance and the eccentricity are exaggerated in the figure. The actual eccentricity would equal one-half the clearance between the actual parts. The center of the hub would follow a circular path around the center of the shaft, the radius of which would equal the eccentricity. This condition could also be caused by using a tapered key. It should not be concluded that these arrangements are a poor design in every case. The designer must decide whether eccentricity of this nature is objectionable and, if it is not, the amount that can be tolerated.

In situations that involve a fit, a designer should ask himself the real purpose of the hole, for it may be that the hole is not critical. For example, the spacer in Fig. 8-12 must have the ends parallel, but the hole need not be held to a close tolerance on its diameter; also, considerable tolerance could be allowed on the squareness of the diameter with the ends.

There are tables that are recognized as standard that give values to be used in determining the limits for both the shaft and the hole for various types of fits. They should be used whenever possible, but only after careful consideration of all the factors involved, never just copied blindly.

8-9 THE DESIGN OF INTERFERENCE FITS

The amount of interference required varies with the purpose of the fit. A very small amount of interference is used to avoid some of the difficulties accompanying clearance that were discussed in Sec. 8-7. A small interference is used for long fits, or for fits involving thin sections. Larger amounts are used where permanent assembly is intended, or where the fit is expected to transmit a force.

When the amount of interference is large enough to require a force

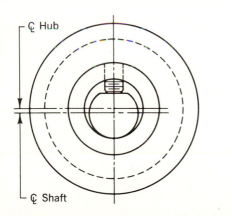
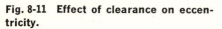
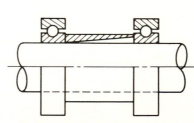

Fig. 8-11 Effect of clearance on eccentricity.

Fig. 8-12 A noncritical hole.

greater than that which can be produced by the equipment available, or where the design is such that the force cannot be applied, a shrink fit may be used. The term "shrink fit" refers to a method of producing an interference fit without force. It is not a certain amount of interference. The outer part is heated — the inner part may also be cooled — to increase the size of the hole to the extent that the parts can be assembled. When the parts return to the temperature at which the interference was measured, a fit of that amount of interference will exist. A medium fit should have an average interference of about 0.0005 in. for each in. of hole diameter. Half this amount is appropriate for fits that require a small interference, and twice this amount is suitable for fits that require a larger interference.

When designing the fit, consideration should be given to the tensile stress induced in the part that has the hole. This stress is approximated with the equation

$$s_t = \frac{Ea}{D} \qquad \text{psi} \tag{8-1}$$

where s_t = tensile stress, psi

E = modulus of elasticity, psi

a = interference, in.

D = nominal diameter of hole, in.

This stress must be less than the yield strength in tension, but due to the nature of the stress it need not be appreciably less.

The temperature required for a shrink fit should be checked by the designer to determine if it is high enough to cause the part to be damaged. This temperature can be calculated with the equation

$$T = \frac{a}{D\alpha} \tag{8-2}$$

where T = increase in temperature, °F

α = coefficient of expansion, in./(in.) (°F)

The interference or the temperature should be increased slightly to allow for cooling and to facilitate the assembly.

Example 8-2 Determine the limits for the shaft and hub, the stress in the hub, and the temperature required for a shrink fit intended as a permanent assembly, where the nominal diameter of the shaft is 4 in., and both shaft and hub are steel.

Solution For a permanent assembly, an appropriate average interference is 0.001 in. per in. of diameter, or 0.004. The hole will have to be bored and the shaft ground; therefore, reasonable tolerance for each is 0.0005. Make the hole 4.0000/4.0005. The shaft should then be 4.0045/4.0040. The maximum interference is

0.0045, and the minimum is 0.0035. The average is the 0.004 required. The stress is found using Eq. (8-1):

$$s_t = \frac{Ea}{D} = \frac{30,000,000 \times 0.0045}{4} = 33,800 \text{ psi}$$

The maximum interference is used to give the greatest stress. Any steel will have sufficient strength, and some cast irons could be used.

The temperature is found using Eq. (8-2):

$$T = \frac{a}{D\alpha} = \frac{0.006}{4 \times 0.000007} = 214°F$$

The 0.006 includes an allowance of 0.0015 to allow for cooling and to facilitate assembly. The 0.000007 is the coefficient of expansion for steel. Assuming a room temperature of 70°F, the temperature of the part need not be greater than 284°F, which will not damage the part.

Interference fits may be used to avoid the inaccuracies caused by clearance, but they can also be the source of inaccuracies. When a single keyway is used, the center of the hub is displaced relative to the center of the shaft due to the fact that there is less metal on the side with the keyway resisting the pressure caused by the interference. This situation can be corrected by using two keys, or by machining a flat on the side opposite the keyway equal in width to the keyway.

When a thin-walled tubular part is pressed into a more substantial part, the inside diameter of the inner part is reduced by an amount approximately equal to the interference. This reduction can be allowed for in making the parts, or if the design permits, by machining the inside diameter after assembly. If only a portion of a thin-walled part is pressed into another part, the condition shown in Fig. 8-13 results. This situation cannot be allowed for in machining the parts, and the inside diameter must be machined after assembly.

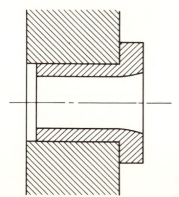

Fig. 8-13 Effect of press fit on a part with a thin wall.

The magnitude of these undesirable effects increases with an increase in the amount of interference. Therefore, to avoid the inaccuracies that may be caused by interference fits, the interference should be kept to a minimum and the parts should be symmetrical in the area of the fit.

8-10 SELECTIVE ASSEMBLY

When the difference between the minimum and the maximum clearance or interference must be small, the tolerances on the mating parts must be greatly reduced. This may result in an excessive cost. A procedure known as selective assembly may be used to avoid this cost when a large number of assemblies are to be produced. It consists of producing parts with reasonable tolerances and then sorting the parts into several groups on the basis of size. Assembly is then performed using large shafts with large holes and small shafts with small holes, and so on, for as many groups as are necessary to get the fit desired with the tolerances desired.

There are several disadvantages associated with selective assembly. The sorting operation increases the cost of production. The parts are no longer interchangeable — an assembly rather than the individual parts must be used when replacement of a part becomes necessary. There will be parts for which there is no mating part and they, therefore, cannot be used. Nevertheless, in many instances, selective assembly is preferred to the alternative of excessively small tolerances.

8-11 TOLERANCE PRODUCED BY PROCESSES

The tolerance that can be expected of a manufacturing process is covered in Chap. 4. It will, however, be helpful in this study of tolerance

TABLE 8-3 Typical Tolerances for Manufacturing Processes

PROCESS	TOLERANCE
Turning	0.003
Milling	0.004
Shaping	0.005
Planing	0.005
Drilling (0.5 dia.)	0.008
Reaming (0.5 dia.)	0.0015
Boring	0.003
Grinding	0.001
Broaching	0.002
Honing	0.0005
Lapping	0.0001

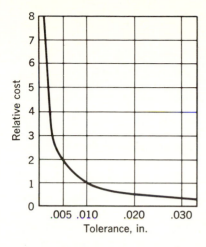

Fig. 8-14 Relation between tolerance and cost.

to have a value that may be regarded as representative for the more common machining processes. This is presented in Table 8-3 for an average-size part. It should be kept in mind that the tolerance which can be held depends on a number of factors, such as: machine condition, material being machined, material of the cutting tool, operator skill, and tooling quality. The values in the table are those that can be expected in production with the exercise of the usual degree of care.

8-12 TOLERANCE AND COST

The tolerance allowed on the dimensions of a part are an important factor in determining the cost of the part. An idea of the magnitude of this factor can be gained from Fig. 8-14, which shows the approximate relationship between machining cost and tolerance.

Some dimensions must be held to close tolerances if the design is to perform its function. These dimensions are usually carefully considered. However, the majority of the dimensions, because they are not critical, are often not given the consideration justified by their effect on cost.

8-13 SURFACE ROUGHNESS

For many parts, the degree of roughness of the various surfaces has a great effect on the proper functioning of the part. The friction between surfaces, whether lubricated or not, increases with an increase in roughness of the surfaces. The wear of a surface that is initially rough is greater than one that is smooth. The performance of an interference fit improves as the surfaces of the mating parts are made smoother. The fatigue strength of

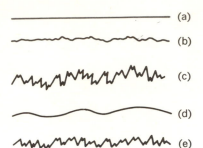

Fig. 8-15 Components of surface roughness.

heat-treated parts is greatly increased with a decrease in surface roughness. Though in most instances as smooth a surface as can be economically produced is desired, there are applications where the surface can be too smooth. Brake drums are such an application.

As was the case with tolerances, the designer is best qualified to determine the surface roughness to be specified.

To describe accurately what is referred to as surface roughness requires several dimensions. There are also a number of terms that have a precise meaning. The side view of a surface, as it appears to the unaided eye, is shown in Fig. 8-15a. A small portion of it, magnified slightly, is shown in Fig. 8-15b. And the same portion, with only the vertical dimensions magnified, is shown in Fig. 8-15c. When carefully analyzed, the surface as represented in Fig. 8-15c is found to consist of several distinct components. In Fig. 8-15d only the wavelike component is shown; this is called waviness. In Fig. 8-15e only the small peaks and valleys are shown. This is technically known as roughness. Thus, it is seen that a surface consists of small irregularities superimposed on large variations similar to gently rolling hills that have just been plowed.

Roughness is due to the action of the cutting tool or grinding wheel, and waviness is caused mainly by such factors as vibration and deflection of the machine or the part. Waviness height, that is, the distance from the top of the wave to the bottom, is measured in inches. Roughness height is measured in microinches, which is millionths of an inch. The maximum peak-to-valley distance can be used, but more often a form of average is used. Actually, there are two methods of arriving at this average. The arithmetical average consists of taking ordinates at equidistant points along the profile, adding them, and dividing by the number of terms. In the root mean square (rms) method, the ordinates are squared, added, and divided by the number of terms, and then the square root is taken. These two methods do not produce identical numerical values for the same surface. The rms value is slightly larger, but the difference is considerably less than the tolerance allowed for the specimens used for determining surface roughness by comparison.

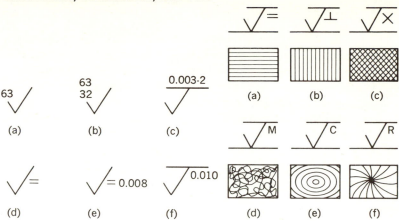

Fig. 8-16 Specification of surface roughness.

Fig. 8-17 Interpretation of lay symbols.

If only the maximum roughness is of importance, it is specified as shown in Fig. 8-16a relative to the surface roughness symbol. If both the maximum and minimum roughnesses are limited, this is indicated as in Fig. 8-16b. When waviness is important enough to justify its being given consideration in manufacture and inspection, it is specified as shown in Fig. 8-16c. Only the maximum height (0.003 in.) may be specified, or the height in a specified width (2 in.) may be given. If the direction of the small ridges and valleys, known as lay, is important to the proper function of the part, it is indicated as shown in Fig. 8-16d. The interpretation of the symbols used is given in Fig. 8-17. In Fig. 8-17a the lay is parallel to the line to which the symbol is applied; in Fig. 8-17b it is perpendicular; in Fig. 8-17c it is angular in both directions; in Fig. 8-17d it is multidirectional; in Fig. 8-17e the pattern is approximately circular relative to the center of the surface; and in Fig. 8-17f it is approximately radial relative to the center. If the roughness width, that is, the distance from one peak to the next, is to be limited, the maximum in inches is placed to the right of the lay symbol, as shown in Fig. 8-16e. When it is necessary to indicate the roughness width cutoff, it is placed as shown in Fig. 8-16f. This number is the distance over which the surface roughness is to be measured and is used to limit the effect that waviness has on the roughness measurement. Standard distances are (in inches): 0.003, 0.010, 0.030, 0.100, 0.300, and 1.000. The distance 0.030 is preferred.

8-14 TYPICAL SURFACE ROUGHNESS VALUES

The surface roughness produced by a process varies considerably. For example, the normal range for rough turning is 1000–500 μin.; for finish turning it is 125–32 μin. However, with extra effort and better than average equipment, 16 or even 8 may be produced in some instances. Thus, the range for turning would be from over 1000 to 8. For grinding, the range is even greater. The material also has an effect on the surface roughness produced by a particular process. A smoother surface is possible with hardened steel than with soft steel. Even the material from which the cutting tool is made affects the surface roughness that is produced.

The surface roughness produced by the various manufacturing processes is covered in Chap. 4. However, an approximate value that may be regarded as typical for a process would be helpful in developing a working concept of surface roughness. In Table 8-4 a roughness-height value is given for various processes. This value is intended as characteristic of the process, and is what can be expected in production from an average machine as a minimum when no special effort is exercised. Only finish operations are shown.

The surface roughness that may be permitted for a particular application cannot be determined by calculation. The effect of surface roughness and an appropriate value in a particular situation will be discussed in the following chapters. Nevertheless, typical applications for various values would aid in developing the working concept of surface roughness, which is the purpose of this section and Sec. 8-15. In Table 8-5 such typical applications are given.

8-15 SURFACE ROUGHNESS AS RELATED TO TOLERANCE AND COST

Comparison of the roughness height with dimensions will aid in visualizing the magnitudes involved in surface roughness. A roughness

TABLE 8-4 Typical Surface Roughness of Manufacturing Processes

ROUGHNESS, μ IN.	PROCESS
	Machined surfaces
125	Shaping, planing, drilling
63	Milling, turning, boring
32	Broaching, reaming
16	Grinding
8	Honing
2	Superfinishing, lapping
	Natural surfaces
500	Sand cast
250	Forged
63	Die cast
32	Cold drawn

TABLE 8-5 Typical Applications of Surface Roughness

250	Clearance surfaces. Maximum for final machining for quality machine work except for large rough parts.
125	Mating surfaces where there is little or no motion between them and where bearing loads are light. Maximum for final machining after heat-treating for parts heat-treated to 180,000 psi or over.
63	Press fits. Rolling surfaces such as cams with rolling followers. Gear-locating faces. Bearings where the motion is intermittent and the load is moderate.
32	Gear teeth. Sliding or rotating bearings with moderate to heavy loading.
16	Precision journal bearings. Gear teeth subject to heavy loading. Friction surfaces such as clutch plates.
8	Surfaces for hydraulic dynamic seals. Crank pins and production gages.
4 and less	Primarily precision measuring instruments. These values should be specified only after thorough investigation has established that they are necessary.

height of 100 μin. will have a distance from a high peak to a deep valley of 0.0005 in., and a distance from one peak to the next between 0.010 in. and 0.100 in., depending on the process that produced the surface. Thus, the

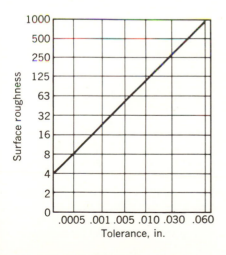

Fig. 8-18 Relation between surface roughness and tolerance.

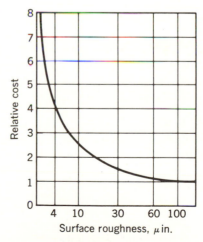

Fig. 8-19 Relation between surface roughness and cost.

surface is actually more like rolling hills than like the jagged mountains appearing in profiles that have the vertical distance exaggerated.

There is a relationship between surface roughness and tolerance such that, in the effort to maintain a close tolerance, a smooth surface is produced as a by-product. Also, when a smooth surface is produced by most of the processes that are used to produce smooth surfaces, it requires little more effort to hold a close tolerance also. It would be meaningless to require, for example, a tolerance of 0.0002 in. between surfaces whose roughness height was 250 μin., because the peak-to-valley distance on one of the surfaces would be 5 times as great as the tolerance. The approximate relation between surface roughness and tolerance is shown in Fig. 8-18.

A typical manufacturing sequence required to produce an exceptionally smooth surface is: rough machine, finish machine, heat-treat, rough grind, finish grind, and then lap, hone, or superfinish. The number of different processes involved, the cost of the machines, the time required to perform the processes, and the care in handling explain the greater cost of producing smooth surfaces. The approximate relation of surface roughness to cost is shown in Fig. 8-19. Again, this is intended only to aid in developing a concept of surface roughness.

QUESTIONS

8-1 What is the difference between functional and nonfunctional surfaces?

8-2 Explain what is meant by the second-degree relationship between surfaces.

8-3 What is the meaning of the term tolerance?

8-4 Explain geometric tolerances.

8-5 Explain true-position dimensioning.

8-6 What is a tolerance stack?

8-7 What effect does a clearance fit have on eccentricity?

8-8 What are the advantages and disadvantages of selective assembly?

8-9 Why is a designer concerned with surface roughness?

8-10 In general terms, compare a surface whose roughness is 125 with one that is 32.

PROBLEMS

Note: It is *recommended* that the introduction to Chap. 19 be read and Projects 1 and 2 be completed after solving some of these problems.

8-1 A shaft has a diameter of 1.495/1.490, and it is placed in a hole that has a diameter of 1.499/1.500. What is the fit?* (An asterisk indicates a problem for which an answer is given.)

8-2 A shaft has a diameter of 2.994/2.992, and it is placed in a hole that has a diameter of 2.498/2.501. What is the fit?

8-3 A shaft with a diameter of 2.3752/2.3748 is placed in a 2.3750/2.3755 hole. What is the fit?*

8-4 A shaft with a diameter of 1.502/1.501 is placed in a hole with a 1.499/1.501 diameter. What is the fit?

8-5 A shaft is to be used in a 0.625-diameter hole that is reamed to ±0.0005. The minimum clearance must be 0.0002, the maximum 0.002. What is the diameter of the shaft?*

8-6 A pin of 0.5005 ± 0.0002 diameter is to be pressed into a hole. The maximum interference is to be 0.0015, the minimum 0.0005. What is the diameter of the hole?

8-7 What tensile strength is induced in a hub made of steel with a modulus of elasticity of 30,000,000 by pressing into it a 2-in.-diameter shaft, if the maximum interference is 0.005?*

8-8 What tensile strength is induced in a hub made of steel with a modulus of elasticity of 29,000,000 by pressing into it a 3.50-in.-diameter shaft, if the maximum interference is 0.009?

8-9 What increase in temperature is required to permit the parts of Prob. 8-6 to be assembled as a shrink fit?*

8-10 What increase in temperature is required to permit the parts in Prob. 8-8 to be assembled as a shrink fit?

SELECTED REFERENCES

ASA B4.1: "Preferred Limits and Fits for Cylindrical Parts," American Standards Association, New York, 1955.

ASA B46.1: "Surface Texture," American Standards Association, New York, 1962.

ASA Y14.5: "Dimensioning and Notes," American Standards Association, New York, 1957.

9 FASTENERS

9-1 INTRODUCTION

An efficient and economical design depends not only on the various components, but on the suitability of the means by which they are joined. This chapter is concerned with fasteners that permit disassembly.

There are several reasons for incorporating a fastener: to permit disassembly for inspection, maintenance, or repair; to permit the product to be made up of smaller units which are easier to manufacture and ship; to provide for the adjustment required to allow for tolerance and wear.

The selection of a fastener requires consideration of a number of factors:

1. The frequency of disassembly.
2. Service conditions such as vibration, impact, and corrosion.
3. The materials to be joined.
4. The function to be provided; that is, will the fastener serve both to join and to assure proper location?
5. Is more than one type of fastener to be used in the joint?
6. Whether a few large or many small fasteners is most appropriate.

There are over 500,000 standard fasteners available. Thus, this discussion must be limited to principles and a description of the more common types.

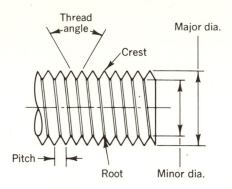

Fig. 9-1 Screw thread terminology.

9-2 THREADED FASTENER TERMINOLOGY

The thread on a bolt is called an external or male thread, and that in a nut an internal or female thread. A thread may be either right-hand or left-hand. A right-hand bolt advances into an internal thread when turned clockwise. A right-hand bolt placed horizontally has a thread that slants down to the right. It is always understood that a thread is right-hand unless otherwise specified.

Figure 9-1 illustrates some of the terms pertaining to screw threads. The major diameter is the largest diameter of the thread and applies to either an external or an internal thread. It is also the nominal size of the thread. The minor diameter is the smallest diameter and applies to either external or internal threads. For external threads, the minor diameter is also called the root diameter. Pitch is the distance from a point on a thread to a corresponding point on the next thread. It is equal in inches to 1 divided by the number of threads in 1 in.

For a single thread, the pitch and lead are equal and a bolt will advance this amount when turned one revolution. Most threads are single, but occasionally a multiple thread is used. A multiple thread is also referred to as a multiple-start thread. A double or two-start thread is really two threads, the lead being twice the pitch, and when it is turned one revolution it will advance a distance equal to the lead. A triple or three-start thread has a lead equal to 3 times the pitch.

9-3 THREAD FORMS

Several thread forms are shown in Fig. 9-2. The acme and square forms are not used for fasteners. They will be discussed in Chap. 18. The American National form was the standard in the United States, and the Whitworth was the British standard. The Unified thread form shown in Fig. 9-3 is now the standard in the United States, Canada, and Great Britain. The Unified

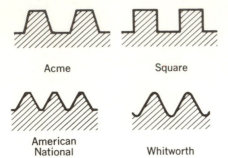

Acme   Square

American
National   Whitworth   **Fig. 9-2   Thread forms.**

form is used for practically all threaded fasteners in the United States and, therefore, will be the only one discussed further. It should be mentioned here that there is so little difference between the American National and the Unified that they may be used interchangeably.

Threads are classified into various series on the basis of the number of threads per inch relative to the diameter. Seven series cover most of the threads for fasteners used in mechanical design. The coarse-thread series, identified as UNC, should be used in all applications where a finer thread is not required for a specific purpose such as to provide fine adjustments. This series has greater resistance to stripping and is less sensitive to damage than the finer threads. This series is particularly recommended for use in soft metals and plastics. The fine-thread series (UNF) is used in applications

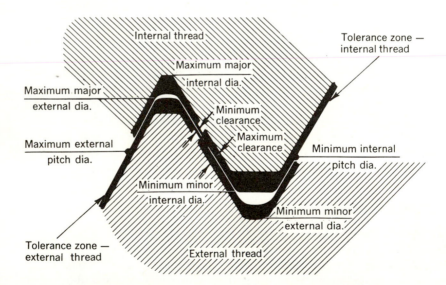

Fig. 9-3   Unified thread form.

where vibration is an important factor, as fine threads are easier to tighten and are less likely to loosen. The extra-fine series (UNEF) is used where thin walls are involved and where the largest number of threads is required for a given length of engagement. In the three series above, the number of threads per inch varies with the diameter, as shown in Table 9-1. In the four remaining series, the number of threads per inch does not vary with the diameter. Thus, all diameters in the 4-thread series have 4 threads/in. The 4-thread series (4 UN) is a continuation of the coarse-thread series in larger sizes. The 12-thread series (12 UN) is a continuation of the fine-thread series, and the 16-thread series (16 UN) is a continuation of the extra-fine series for large-diameter threads. The 8-thread series (8 UN) is often used in applications where the coarse-thread series would have fewer than eight threads per inch.

An effort should be made to use only threads designated as Unified, and care should be taken to specify only standard threads.

A tolerance is required in the manufacture of threads, and a fit exists in a manner similar to a shaft and hole. The threads of a given series are divided into classes on the basis of the fit between the internal and external thread. The classes are designated 1, 2, 3, and 5. In addition, an external thread is identified by the letter A, and the internal thread by the letter B.

TABLE 9-1 Unified Screw Threads

	THREADS PER INCH		
SIZE	UNC	UNF	UNEF
¼	20	28	32*
⁵⁄₁₆	18	24	32*
⅜	16	24	32*
⁷⁄₁₆	14	20	28
½	13	20	28
⁹⁄₁₆	12	18	24*
⅝	11	18	24*
¾	10	16	20
⅞	9	14	20
1	8	12	20
1⅛	7	12	18*
1¼	7	12	18*
1⅜	6	12	18*
1½	6	12	18*
1¾	5		16
2	4½		

*An American National thread.

Thus, a ¾-in.-diameter external class 2 fine thread would be specified as ¾-16 UNF-2A. A corresponding nut would be ¾-16 UNF-2B.

The allowance for a thread is the difference between the maximum external pitch diameter (PD) and the minimum internal PD. A class 1 thread has an allowance and considerable tolerance; a class 2 thread has an allowance but less tolerance; a class 3 thread has no allowance and less tolerance than class 2; class 5 is an interference fit.

Class 1 is used where parts must assemble readily even if the threads are slightly damaged or dirty. Class 2 is the most often specified class. The allowance minimizes the possibility of galling and permits plating or other coatings. Class 3 is used only where a close fit or extreme accuracy are essential. The class 5 fit is used primarily for studs. The clearances and tolerances for a class 2 thread are indicated in Fig. 9-3. An internal thread of one class may be used with an external thread of another class.

9-4 BOLTS AND STUDS

Standard bolts are available in many forms for specific purposes. The most common has a hexagon head. The regular hexagon bolt is not finished on any surface; the semifinished hexagon bolt is finished to produce a flat surface under the head; the finished hexagon bolt is finished all over. However, the term "finished" in this case does not mean that the surface is machined, but it implies only a higher quality and closer tolerances in manufacture.

Regular and semifinished bolts are standard in sizes from ¼ in. in diameter to 4 in. in diameter. The finished bolts are standard from ¼ to 3 in. in diameter.

Also standard are bolts designated as heavy, and they may be regular, semifinished, or finished; for example, a "heavy finished hexagon bolt." The head of a heavy bolt has a larger hexagon and greater thickness. They are available from ½ to 3 in. in diameter.

Bolts are available in a number of different materials. The most common are known as SAE grade 2 and SAE grade 5. The SAE grade 2 is low-carbon steel which has a minimum tensile strength of 69,000 psi for ½-in. diameter and smaller. The SAE grade 5 is medium-carbon steel that is heat-treated, and for ¾-in. diameter and smaller has a minimum tensile strength of 120,000 psi. Larger sizes have lower strength. The SAE grade 5, though more expensive, is more economical where a large clamping force is required.

A stud is a fastening device, threaded on both ends, that is screwed into a thread in one of the parts to be joined; a nut is used on the other end in the usual manner. Studs are used in situations where a bolt and nut cannot be used, and where a bolt alone is not appropriate because of the possibility of damaging the internal threads by frequent removal. A brittle material, such as cast iron, is susceptible to such thread damage. A joint using studs instead of bolts is more expensive, because of the extra operations of tapping the hole and inserting the stud. Therefore, an effort should be made to avoid them.

A stud should be held in the tapped hole by the tightness of the thread not by jamming it against the bottom of the hole. Thus, a class 5 fit is recommended. The thread engagement should be at least 1.25 times the diameter of the thread for steel, 1.5 times for cast iron, and as much as 2.5 times the diameter for other materials.

9-5 NUTS

Nuts are available in a great many different forms. There are several reasons for the great variety: to provide a means of preventing the nut from loosening, to provide an internal thread in a location that becomes inaccessible at assembly, or to reduce the total in-place cost of the internal thread.

Conventional hexagon nuts are available in the following forms: unfinished, as plain or jam, and either regular or heavy; semifinished, as plain, slotted, or jam, and either regular or heavy; finished, as plain, slotted, jam, also as thick plain, thick slotted, and castle. The semifinished nuts are finished on the bearing surface. The term "finished" has the same meaning that it did for bolts. Heavy nuts are ⅛ in. larger across the flats of the hexagon. Slotted nuts have slots for cotter pins. Jam nuts are thin and intended as a locking device.

When a nut is tightened on an external thread there are tendencies: for the threads in the nut to shear out; for the major diameter of the threads in the nut to increase, called dilation; and for the nut to crush the part it bears against. The regular nuts mentioned above are designed to prevent trouble from these three tendencies in usual situations. The thick nut has greater resistance to the thread's shearing. The heavy nut has less tendency toward dilation or crushing of the part it bears against.

A heat-treated nut will have greater strength. Nevertheless, use of a soft nut is usually preferable because it will more readily deform to compensate for variations in the threads, and it thus distributes the load more evenly.

If a nut is tightened so as to produce a high tensile stress in the bolt, and if the bolt has sufficient length, a means of locking the nut is unnecessary, even in an application involving vibration. There are, however, many situations where these requirements cannot be met, and some means of locking the nut is necessary. A conventional plain nut may be locked by deforming the external thread that extends beyond the nut, or by deforming the threads of the nut after it is tightened. These methods are undesirable if disassembly is required. Special washers can be used, and these will be discussed later. Another method is to use two nuts; the jam nut was designed for this purpose. A jam nut should be avoided where tension is to be developed in the bolt by tightening the nut.

A slotted nut, with a cotter pin to prevent the nut from coming off, is often used, but it has the disadvantage of having to be backed off or tightened from the desired position to permit inserting the cotter pin.

Self-locking nuts, or locknuts as they are often called, may be placed in one of two broad classifications: prevailing-torque or free-spinning. Free-spinning locknuts turn freely until they become seated. Further tightening produces the locking action. This type should be avoided where there is relative motion between the parts joined, or where the tension in the bolt may be decreased by embedding the bolt head or the nut or by deformation of the parts. Prevailing-torque locknuts turn freely for the first few turns and then engage the locking feature. They must be wrenched to final position. This type should be avoided where the nut must travel a considerable distance on the thread to its final position, because of the possible damage to the locking feature, or where there is relative motion between the joined parts that tend to loosen the nut. Those with nonmetallic inserts should not be used for temperatures above 250°F.

There are several types of nuts that may be permanently attached in the proper location before assembly, thus permitting the use of a nut in locations where it is impossible to place a wrench when the parts are being assembled. This type has a high installed cost and should be used only where one of the more common nuts cannot be used.

9-6 SCREWS

The more common screws fit into one of the following categories: machine screws, cap screws, wood screws, setscrews, and self-tapping screws. Standard machine screws are available with a number of different head forms. They are made with the regular screwdriver slot and several styles of driving recess. Screws 2 in. long and shorter are threaded for their full length. Either coarse or fine class 2A threads are available. The diameter may be as large as ¾ in. for some types. Machine screws are generally supplied with flat points.

There are fewer head forms for standard cap screws, though they may be obtained with hexagonal-socket head or fluted-socket head, in addition to the usual driving means. The thread length is equal to twice the diameter plus ¼ in. The points are chamfered. Either fine or coarse class 3A threads are available. The maximum diameter for some types is 1½ in.

A socket-head screw has a small area under the head; consequently the

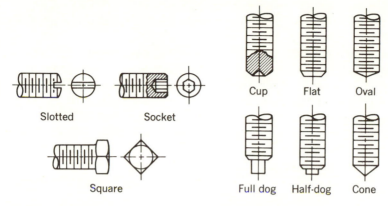

Fig. 9-4   Forms of setscrews.   **Fig. 9-5   Setscrew points.**

bearing stress is high. Unless the part against which it bears is heat-treated, crushing will take place before the full tensile strength of the screw is reached.

A shoulder screw has a socket head and a thread that is smaller than the body of the screw. It is made from alloy steel and is heat-treated. The body is ground to a close tolerance. Shoulder screws are often used as pivot pins in linkages.

Wood screws are available in sizes from No. 0 to No. 24. The size refers to the diameter of the shank under the head and is about $\frac{1}{16}$ in. for the No. 0 and $\frac{3}{8}$ in. for the No. 24. For larger sizes, a lag bolt is used.

Setscrews are short screws, the primary purpose of which is to secure a hub to a shaft and transmit a relatively small torque. The forms available are slotted, socket, and square, as shown in Fig. 9-4. For the slotted type, the slot is deformed at a torque much less than that required to strip the threads. The square-head type can be tightened to a much greater extent but should not be used on moving parts where they could catch the clothes of workmen. In some states their use in such an application is prohibited by law.

There are six standard points, each of which is available in each of the four forms. They are shown in Fig. 9-5. The cup point is most often used, though the burr that it produces may interfere with removing the hub. The flat point does little damage to the shaft and can be used where fine adjustments are required. The dog points have greater holding power when a hole is drilled in the shaft to match the point diameter.

The diameter of the setscrew should be about one-fourth the diameter of the shaft. If more than one is to be used, it should be placed next to the first along the shaft. If the second must be placed in the same location along the shaft as the first, but at an angle to it, the angle should be about 60°.

Tapping screws are designed to provide their own mating thread as they are used at assembly; thus, a considerable saving of time is achieved. Also, because the screw makes its own internal thread, there is no clearance; thus, a tight fit is produced.

There are two basically different types of tapping screws: the thread-forming type and the thread-cutting type. The thread-forming type produces a thread in a manner similar to the process of thread rolling, while the thread-cutting type produces a thread in a manner similar to tapping. Either type may produce a thread that is similar to a standard thread, or a thread that has about twice the pitch of a standard thread, known as a spaced thread.

There are several methods of producing the thread employed by the thread-forming type. There are also several styles of points that cut a thread. The type to use depends mainly on the material in which the threads are to be produced, though the fact that no chips result from the use of the thread-forming type may be an important factor in some applications. In general, the thread-forming type should be used in materials that are ductile enough to stand the deformation involved, and for brittle materials the thread-cutting type should be used. The spaced-thread types drive faster and are best in material that tends to crumble, but the types that produce a thread similar to a standard thread have a greater clamping force and greater resistance to loosening. A large variety of standard head styles are available. The recommendation of the manufacturer of a particular screw should be followed in specifying the size of the hole in which the threads are to be produced.

9-7 STRENGTH OF THREADED FASTENERS

For convenience, in this section the member of a threaded connection that has an external thread will be referred to as a bolt, and the member with an internal thread as a nut. Sources for the stress that may be imposed on a bolt or nut are the external load encountered in service, the load developed to make the joint tight, and the load involved in the tightening process. The stresses that these loads produce are as follows:

1. A bearing stress caused by the threads of the bolt bearing on the threads of the nut. This stress is seldom critical but may result in galling.
2. A shear stress in the threads, that is, a tendency to strip the threads. A full-depth thread has the minor diameter of the bolt and nut nearly equal. The standard Unified thread is based on a 75 percent thread depth. A regular nut drilled out to only 50 percent of a full-depth thread, when tightened, will break the bolt rather than strip the threads of the nut. A standard thread will develop the full tensile strength of the bolt when engaged a distance of about one-half the diameter of the bolt. The thickness of

a regular nut is almost equal to the diameter of the bolt. Thus, the shear stress in the threads need not cause trouble.

3. A shear stress in the body of the bolt when the bolt is used in shear. This becomes a simple problem of selecting a bolt that has sufficient area in shear.

4. A tensile stress in the bolt. If the bolt is used in an application such that there is no load due to tightening and only a static applied load, the problem is merely one of selecting a bolt with adequate area at the root diameter of the thread. If the load is not static, the problem appears more complex. However, if the bolt is tightened to the extent that a tensile stress is developed that is almost equal to the yield strength, any load that induces a stress less than this will have no effect on the bolt. Tightening a bolt in this manner is called preloading, and a bolt that is preloaded to this extent is immune to fatigue. Also, a bolt that is properly preloaded will not loosen unless the preload is reduced in some manner, such as crushing of the part under the bolt head.

5. A torsion stress in the bolt due to tightening. As a bolt is tightened the tensile load that is induced forces the threads of the bolt and nut together, increasing the friction. The friction increases with an increase in the tensile load. A torque must be applied to overcome this friction and to develop the tension load. When the tightening is completed, the torsion stress disappears; thus, the stress in the bolt is reduced. Therefore, if the joint is properly designed and the bolt does not fail while it is being tightened, it should not fail in service.

The torque required to produce a given tension load can be approximated with the equation

$$T = CDP \tag{9-1}$$

where T = torque, lb-in.
C = coefficient
D = bolt diameter, in.
P = tension load in bolt, lb

The coefficient C depends on the presence of lubrication. For unlubricated bolts and nuts, it may be taken as 0.2; if they are waxed or lubricated, 0.15 is more appropriate.

9-8 DESIGN OF THREADED PARTS

The detailed design of threaded parts is important from both the manufacturing and strength points of view. An external thread should have a thread relief as shown in Fig. 9-6. The width should be at least equal to $2\frac{1}{2}$ times the pitch of the thread, and the maximum diameter should be less than

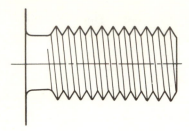

Fig. 9-6   Thread details.

the minimum minor diameter of the thread. The radius should be one-half the difference between the major and minor thread diameters. The end should be chamfered 45° to a diameter less than the minimum minor diameter of the thread, as shown in Fig. 9-6.

Internal threads for a steel bolt or screw subject to a tension load should provide an engagement in steel equal to the thread diameter, 1½ times the thread diameter for cast iron or brass, and 2 times the thread diameter for aluminum or zinc. If the threads are to be tapped in a blind hole, the depth of the hole should be deeper than the depth of full thread required by a distance at least equal to 5 times the pitch of the thread. If a part with an external thread must seat against the bottom of a threaded hole, a thread relief should be provided in the hole; the width should be equal to 3 times the thread pitch, and the diameter should be greater than the maximum major diameter.

When a bolt is subjected to a suddenly applied load that tends to lengthen it, energy must be absorbed. To absorb energy there must be a deformation. A bolt that has the diameter of the body reduced to slightly less than the minimum minor diameter of the thread will have the length of the body deformed by the load. If the body diameter is not reduced, the portion of the thread between the nut and the body will have to take most of the required deformation, and thus the stress will be greatly increased. The bolt may be made as shown in Fig. 9-7 to maintain alignment of the joined parts.

If position, alignment, or concentricity is important, the thread must

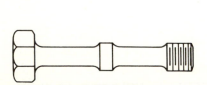

Fig. 9-7   Bolt for a shock load.

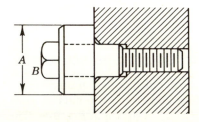

Fig. 9-8   Bolt with pilot.

not be relied upon to provide it. A pilot diameter, such as that shown in Fig. 9-8, should be used. *A* is the diameter that must be accurately located, and *B* is a square for a wrench.

The meaning of the class of a thread must be kept in mind. A class 3 thread is manufactured to a closer tolerance than is a class 2. Thus, a class 3 costs more and should not be specified unless the closer fit or closer tolerance is necessary.

9-9 WASHERS

Washers perform a number of functions when used with other fasteners. The familiar flat washers are known as plain washers. They are used to increase the bearing area under a nut or the head of a screw or bolt and should be used with soft metals. When oversize holes are used to facilitate assembly or to allow for greater tolerance in the location of the holes, a plain washer is required. If a slotted nut and cotter pin are used, one or more plain washers may be necessary to bring the nut to the proper position relative to the hole in the bolt.

Washers of various designs are intended to keep a threaded fastener from loosening. There are two methods used: one is to provide a spring effect that is supposed to maintain the tension in the bolt or screw; the other is to provide teeth that dig into the fastener and the adjacent part and thus prevent the fastener from loosening. Often, both methods are incorporated. A great many different types are available, and most of them are quite effective. However, where only one fastener is involved in a critical application, or the application does not permit preload, a more positive means of locking is desirable. A slotted nut and cotter pin, it will be remembered, has the disadvantage that the nut may have to be rotated from the desired position to permit installing the cotter pin. Washers similar to those shown in Fig. 9-9 may be used to provide positive locking at the position that gives

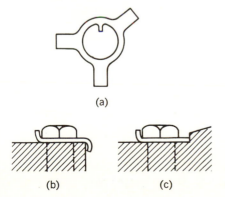

(a)

(b)   (c)   **Fig. 9-9 Locking washers.**

the required fit by bending a tab up against the flat of the nut or bolt head. The type with the internal tongue, Fig. 9-9a, requires a keyway in the thread. The forms shown in Fig. 9-9b and c are used with a fastener in a tapped hole; if they are used with a nut, the head of the bolt or screw must also be prevented from turning.

9-10 PIN FASTENERS

Pins can be used effectively in many applications where the fastener is subject to a shear load. The simplest is a dowel pin, which is a plain cylindrical pin with a slight chamfer on one end and a corner radius on the other end. They are hardened and ground to a tolerance of ±0.0001 in. They require a hole reamed to a rather close tolerance that will produce a press fit in at least one of the parts joined.

A taper pin has one end smaller than the other. The taper is 0.250 in./ft on the diameter. Thus, the small end of a taper pin 3 in. long will be 0.0625 in. smaller than the large end. The holes in both parts must be simultaneously reamed with a taper reamer. The taper pin is hammered in, but there can be no press fit. Sometimes the taper pin is prevented from falling out either by bending over the small end, which is made to extend by additional reaming, or by reaming the holes to where the large end is below the surface and the hole is deformed. Taper pins with a thread on the small end are available; a nut is then used to draw the pin into place and keep it there. A special type of washer is also required.

A clevis pin is a simple pin with a small head on one end and a hole for a cotter pin at the other end. It is used where frequent disassembly for adjustment or maintenance is involved. A plain washer should be placed between the cotter pin and the part.

The need for a pin that could be used in a drilled hole but that would remain in place as though it were a press fit has resulted in the development of several unique pin fasteners. They are called radial locking pins. The

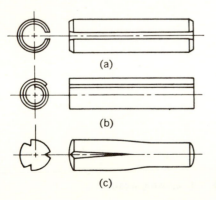

(a)

(b)

(c)   **Fig. 9-10 Radial locking pins.**

slotted tubular type is shown in Fig. 9-10a. These pins are formed from sheet and heat-treated to an extent that they possess springlike characteristics, which keep them tight in a drilled hole. They are suitable for applications where vibration and shock are present. Slotted pins are designed to permit placing a smaller pin inside a larger one to increase the shear strength as well as the fatigue and impact strength. The spiral-wrapped type is shown in Fig. 9-10b. They also have springlike characteristics that compensate for the greater tolerance of a drilled hole. They are produced in three series: light-duty, medium-duty, and heavy-duty. An example of a groove pin is shown in Fig. 9-10c. Groove pins are solid pins that have grooves pressed into the surface which increase the diameter. When pressed in a hole, the increased diameter tends to act as a press fit, but there is not the rigidity of a plain pin. Thus, a range of hole sizes can be accommodated. This range is considerably smaller than that permitted for the other two radial locking pins. The advantage of this type is that the grooves can be placed only at one end or in the middle. This permits their use in applications where the slot or step which are features of the other types would be undesirable. The radial locking pins are available from about $\frac{1}{32}$ to $\frac{1}{2}$ in. in diameter.

9-11 RETAINING RINGS

A retaining ring is, in effect, a shoulder that can be placed on a shaft or in a hole to keep the components of an assembly in their proper position. Some advantages that may be gained by the use of retaining rings are: faster assembly, fewer machining operations, less complex components, fewer parts, and reduced size. The grooves that are required introduce a stress concentration which in some possible applications would be very undesirable.

The two basic kinds of retaining rings are those stamped out of sheet and those formed from wire. They are further classified into three groups:

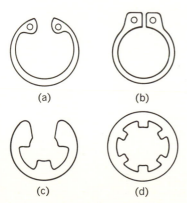

(a)

(b)

(c)

(d)

Fig. 9-11 Retaining rings.

axially assembled, radially assembled, and those that do not require a groove.

Figure 9-11*a* is an example of an axially assembled internal ring. With a pair of special pliers that fit the holes, this ring is compressed and inserted into a hole. Figure 9-11*b* is an example of an axially assembled external ring. It is expanded and slipped over the end of a shaft. Both rings seat in a groove and are thus able to resist a considerable axial load. The design must provide sufficient space for the tool required for assembly. Rings that are bowed or beveled are intended to compensate for tolerances and thus eliminate end play that may be objectionable in some applications.

Figure 9-11*c* is an example of a radially assembled ring. These rings are inserted at 90° to the axis of the shaft. A special tool is not necessary for assembly, but the use of the special tools available greatly reduces the time required.

An example of a ring that does not require a groove is shown in Fig. 9-11*d*. These are pressed over a shaft to the desired position and remain in that position because the prongs grip the shaft. An internal type with the prongs on the outside is also available.

The retaining rings formed from wire are available with a number of cross sections, rectangular and round being the most common. Rings with a rectangular cross section are made in either the internal or external type for axial assembly, also in the external type for radial assembly. The round-cross-section rings are available only in the external type for either axial or radial assembly. The round-cross-section rings are suitable only in applications where the axial load is relatively small.

9-12 MISCELLANEOUS FASTENERS

There are many kinds of fasteners that are not related to those that have been discussed. It would be well to mention some of them.

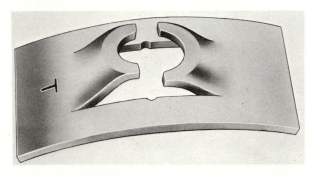

Fig. 9-12 A Speed Nut®. (*Tinnerman Products, Inc.*)

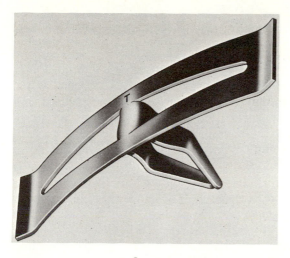

Fig. 9-13 A Speed Clip®. (*Tinnerman Products, Inc.*)

Inserts are internal threads that may be placed in holes for various reasons, such as to provide threads that will not be damaged by frequent disassembly and assembly. They may be screwed into a tapped hole, may tap their own threads, or may be pressed into a hole; or the material of the part may be molded or cast around them.

Nuts stamped from sheet that engage only one thread, shown in Fig. 9-12, are available in various forms for special situations. They have springlike characteristics that resist loosening. Some are so designed that they easily snap into place and remain in the proper position to receive the screw at assembly.

Figure 9-13 shows an example of a spring clip. Fasteners of this kind are designed to accomplish a special function, for example, attaching trim in such a manner that the fasteners are not visible.

There are a number of fasteners the primary advantage of which is that they may be quickly released. They are used to lock securely in place inspection or access panels that must be frequently opened or removed. These fasteners cost more than conventional fasteners and cost more to install. Thus, they should be used only in those situations where it is proved that the time saved in maintenance justifies the extra cost.

Quick-release pins may be obtained that have a means of keeping them in place while subjected to operating conditions, but which can be quickly removed by merely pulling them out or pressing a plunger to permit their removal. They may be inserted just as easily. However, the remarks on the cost of quick-release fasteners made above apply.

9-13 JOINT DESIGN

There is much more to designing a joint than merely selecting a fastener that is strong enough. The parts to be joined must be rigid and must have sufficient compression strength to prevent crushing under the head of the fastener. It is well to keep in mind that a tapped thread is expensive and easily damaged, and when damaged the part may have to be scrapped. Whenever possible, a joint should be designed to put the fasteners in either shear or tension, and they should be placed in the location where they will be most effective. Consideration should also be given to the effect that the holes in the parts being joined will have on the strength of these parts. Another factor to be considered in locating the fasteners is space for both the wrench and the required movement of the wrench — which should be at least 60°.

Bolts subject to shear should be placed with the head in such a position that in the event the nut came off the bolt would tend to remain in place. Bolts and rivets should not be used in the same joint unless the rivets take the entire shear load. This is because bolts do not fill the hole and rivets do, and the rivets would have to be overstressed before the bolts became effective in resisting the shear load. When a joint puts several bolts in shear, the bolts must closely fit the holes in both parts. A threaded fastener must never have the shear plane across the threaded portion.

When a bolt is preloaded to keep it from loosening or to increase its resistance to fatigue, it should be remembered that a long bolt will be more reliable than a short one. The reason for this is that all steel bolts elongate approximately 0.001 in./in. of bolt length for each 30,000 psi of tensile stress; thus, a longer bolt will retain a greater portion of its preload when something causes it to lose a given amount of elongation. This loss may be due to the flattening of a burr, wear or crushing under the head of the bolt or nut, or the chipping away of scale or plating from under the head of the bolt or nut.

If, in a joint subject to repeated loadings, a bolt is of a size such that its tensile strength is 3 times as great as the applied load, the strength of the joint will not be 3 times as strong, unless the bolt is preloaded to 3 times the applied load.

There are several methods of specifying the required preload. The torque to be applied in tightening the nut is often used, but is rather inaccurate because of the variables involved, such as the coefficient of friction. A more precise method would be to specify the amount the bolt must be elongated in the tightening process. In situations where it is impractical to measure the elongation, the amount the nut is turned after all the looseness is taken up can be specified.

Example 9-1   A 0.50-in.-diameter bolt, 6 in. long, with a ½-20 thread, is to be preloaded to a stress in the shank of 80,000 psi. How much must the nut be turned to provide this preload?

Solution   The modulus of elasticity is 30,000,000, and is equal to the stress in psi divided by the strain in inches per inch. Therefore, the strain is equal to the stress divided by the modulus of elasticity; that is

$$\text{Strain} = \frac{80,000}{30,000,000} = 0.00267 \text{ in./in.}$$

The bolt is 6 in. long; therefore, the total elongation is

$$6 \times 0.00267 = 0.016 \text{ in.}$$

There are 20 threads per inch; thus, the pitch is 0.05 in., and one-third of a turn is

$$\frac{0.05}{3} = 0.0167 \text{ in.}$$

Therefore, about ⅓ of a turn, or 120°, will produce the required preload.

The joint shown in Fig. 9-14, known as a knuckle joint, is used to connect two rods which carry a tension load and which must be readily disconnected for maintenance or adjustment. Though there is nothing com-

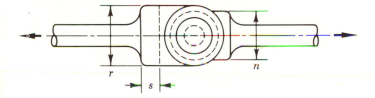

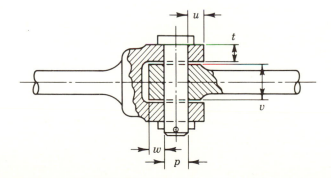

Fig. 9-14   A knuckle joint.

plex about this joint, a detailed description will be given, because it provides an excellent opportunity to illustrate how various stresses are produced in the transmission of a load. This process of load tracing is an essential part of stress analysis.

The load will be thought of as being transmitted from the left rod to the right rod. The area specified will be the area to use in finding the stress (stress equals load divided by area) or the load at which the failure shown in Fig. 9-15 will occur (load equals strength of material times area). To avoid confusion, the total area for the part and the full load acting on the part should always be used.

The rod may fail in tension. The area involved is that of the cross section of the rod. The load divides at the fork end, one half passing into each part of the fork. A shear failure may take place, as shown in Fig. 9-15a; the area is $2rs$. A tension failure may occur, as shown in Fig. 9-15b, known as a tension failure in the net area; the area is $2t(r - p)$. The load is transferred from the fork to the pin. There may be a bearing failure in the fork, as shown in Fig. 9-15c, or in the pin, as shown in Fig. 9-15d; the area in each case is $2pt$. A shear failure may occur, as shown in Fig. 9-15e. This is referred to as tearout, and the area is the distance from the edge of the hole to the edge of the part times the thickness times 2, because there are two areas shearing. For the fork as a whole, the area is $4ut$. The load is transferred to the eye, causing shear stresses in the pin, which could fail in double shear, as shown in Fig. 9-15f. The area is twice the area of the cross section of the pin. The two halves of the load are combined into a single load by the pin. There may be a bearing failure in the eye or in the pin; the area in each

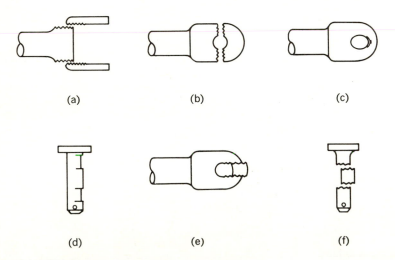

(a)   (b)   (c)

(d)   (e)   (f)

Fig. 9-15 Modes of failure of a knuckle joint.

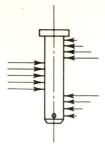

Fig. 9-16 Bending in a knuckle joint pin.

case is pv. The load in the eye may cause a tension failure in the net area, where the area is $v(n-p)$; or it may cause a tearout failure, where the area is $2vw$. The pin is subject to bending. The loading would be about as shown in Fig. 9-16. The maximum bending moment will be at the center of the pin and is equal to $(\text{load}/2)\,(v/4 + t/3)$.

In a well-designed joint, the failures described above would occur at approximately the same load. However, the thickness of the fork t is generally made considerably greater than the theoretical minimum to reduce the tendency of the fork end to spread; that is, to increase the clearance between the fork and the eye. Such spreading would produce stresses for some of the modes of failure in excess of those calculated.

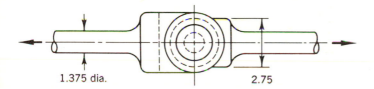

1.375 dia.   2.75

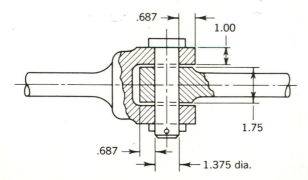

.687 → 1.00

.687 →   1.75

.687 →   ← 1.375 dia.

Fig. 9-17 Example 9-2.

Example 9-2 For the knuckle joint shown in Fig. 9-17 all parts are made from low-carbon steel, the design stresses for which are: tension 18,000 psi and shear 9000 psi. The working load is 25,000 lb. (*a*) Find the stress in the eye for the following modes of failure: tearout, tension in the net area, and bearing. (*b*) Find the working load for the following: pin shear, tearout in the fork.

Solution

(*a*) $\text{Tearout} = \dfrac{25,000}{2 \times 1.75 \times 0.687} = 10,400 \text{ psi}$

$\text{Tension in net area} = \dfrac{25,000}{1.75\,(2.75 - 1.375)} = 10,400 \text{ psi}$

$\text{Bearing} = \dfrac{25,000}{1.375 \times 1.75} = 10,400 \text{ psi}$

(*b*) $\text{Load for pin shear} = 9000 \times 2 \times \dfrac{\pi \times 1.375^2}{4} = 26,700 \text{ lb}$

$\text{Tearout in fork} = 9000 \times 4 \times 0.687 \times 1.00 = 24,700 \text{ lb}$

QUESTIONS

9-1 List four factors that should be considered in selecting a fastener.

9-2 What are the meanings of UNF, UNEF, and 12-UN?

9-3 What is the difference between a class 1 and a class 2 thread?

9-4 How does a regular and a finished bolt differ?

9-5 A heat-treated nut has greater strength, but a soft nut is usually preferred. Why?

9-6 What is the disadvantage of a slotted nut and cotter pin?

9-7 Why should a prevailing-torque locknut be avoided when it is necessary for the nut to travel a great distance on the thread to its final position?

9-8 What is a shoulder screw?

9-9 What are the two basic types of tapping screws?

9-10 Explain the torsion stress induced in a bolt when it is tightened.

9-11 When a bolt is subjected to a suddenly applied load that tends to lengthen it, the diameter of the body is often reduced. Why?

9-12 Describe two types of radial locking pins.

9-13 How is an axially assembled internal retaining ring used?

PROBLEMS

Note: Upon completion of some of these problems, Project 3 in Chap. 19 is appropriate.

9-1 What torque is required to develop a tensile load of 20,000 lb in a ¾-diameter bolt that is not lubricated?*

9-2 What tensile load will a torque of 4000 lb-in. produce in a 1½-diameter bolt that is lubricated?

9-3 A ⅞-diameter bolt, 8 in. long, with a fine thread, is to be preloaded to a stress of 50,000 psi. How much must the nut be turned?*

9-4 The nut on a ⅜-diameter bolt 10 in. long is turned one-half a turn. What stress is developed in the shank of the bolt if: (*a*) the bolt has a UNC thread? (*b*) the bolt has a UNF thread?

9-5 For the situation described in Example 9-2, what is the stress in the fork for the following modes of failure: tearout, tension in the net area, and bearing?

9-6 Explain why the stress due to tension in the net area, tearout, and bearing are identical in Example 9-2.

9-7 A knuckle joint similar to that shown in Fig. 9-16 is subjected to a static design load of 45,000 lb. The pin is to be made of steel heat-treated to 160,000 psi. What is the diameter of the pin?*

9-8 If the fork in the joint shown in Fig. 9-16 is made of steel heat-treated to a tensile strength of 180,000 psi, what is the load at which failure would occur due to tearout?

9-9 A knuckle joint similar to that shown in Fig. 9-16 is subjected to a static working load of 15,000 lb. The eye is made of steel heat-treated to a tensile strength of 140,000 psi, and the pin is 1.500 in. in diameter. The following strength factors apply: $a = 1.2$, $b = 1.5$, $c = 1.0$, and $d = 2.0$. Find the thickness of the eye to provide sufficient strength in bearing.*

9-10 If in Prob. 9-9 the radius of the eye is 1.60, find the thickness of the eye based on: (*a*) tearout and (*b*) tension in the net area.

SELECTED REFERENCE

Laughner, V. H., and A. D. Hargan: "Handbook of Fastening and Joining Metal Parts," McGraw-Hill Book Company, New York, 1956.

10

PERMANENT JOINTS

10-1 INTRODUCTION

Often it is impossible or very inconvenient to assemble a product in the manufacturing plant. Many parts may be made more economically by joining components such as a tube and a small forged end fitting rather than machining a large forging to produce the same configuration. In such instances, a method of permanently joining the members or components is essential. This chapter will deal with the design of the permanent joints most often employed.

Factors to be considered in deciding on a joining method are: the strength required, whether the joint must be fluid tight, whether the joint is to be made in the plant or at the erection site, and the precision required in the relative location of the joined parts.

10-2 RIVETED JOINTS

Some of the advantages of riveting are: joints between dissimilar metals or between parts of different thickness are easy to produce; it is possible to join metals and nonmetals; and much less skill is required than for welding. An important disadvantage is that in most cases it is more expensive than welding. The two most common types of riveted joints are shown in Fig. 10-1. Figure 10-1a is a lap joint and Fig. 10-1b is a butt joint. The short pieces on either side are known as straps or cover plates.

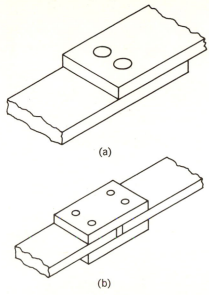

(a)

(b)   **Fig. 10-1 Types of riveted joints.**

10-3 STRENGTH OF RIVETED JOINTS

The strength of a riveted joint involves tensile stress in the net area of the original plates, shear stress in the rivets, and bearing stress in both the rivets and the plates. Failure in tension in the net area is illustrated in Fig. 10-2; the shaded area is the net area. The maximum tensile stress in the net area is found using the equation

$$s_t = \frac{P}{t(w - Dn)} \tag{10-1}$$

where s_t = tensile stress, psi

$\quad P$ = load, lb

$\quad t$ = thickness of the thinner plate, in.

$\quad w$ = width of the plate, in.

$\quad n$ = number of rivets

$\quad D$ = diameter of holes, in.

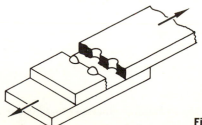

Fig. 10-2 Failure in tension in the net area.

The rivets can be in single shear, as shown in Fig. 10-1*a*, or in double shear, as shown in Fig. 10-1*b*. In double shear there is a tendency to shear the rivet in two places. Thus, the shear stress for single shear is

$$s_s = \frac{P}{nA} \tag{10-2}$$

And the shear stress for double shear is

$$s_s = \frac{P}{2nA} \tag{10-3}$$

where A is the area of the rivet in shear, which is the area of a circle of diameter D.

If the failure was in bearing, the rivet would crush the plate, causing an elongated hole. The rivet could also be crushed if its bearing strength was less than that of the plate, providing it did not shear first. The maximum bearing stress is

$$s_{br} = \frac{P}{nDt} \tag{10-4}$$

where s_{br} = bearing stress, psi
    t = thickness of the thinner sheet of a lap joint or the middle sheet of a butt joint, in.

If the distance between the center of the hole and the edge of the plate, known as edge distance, is not great enough, failure may take place in one of the ways shown in Fig. 10-3.

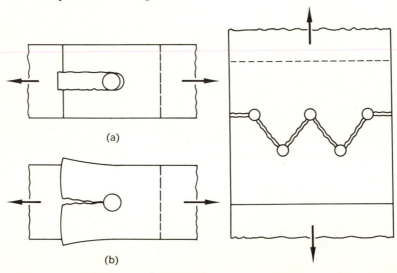

(a)

(b)

Fig. 10-3 Failure due to insufficient edge distance.   **Fig. 10-4 Failure due to insufficient distance between rows.**

Fig. 10-5 The situation for rivets in single shear.

If there are two rows of rivets and the distance between the rows is too small, failure may take place as shown in Fig. 10-4. Prevention of the failures shown in Figs. 10-3 and 10-4 is usually accomplished by adhering to the practices discussed in Sec. 10-4.

The strength and detail design of riveted joints for some applications is governed by codes. The two most common are the American Institute of Steel Construction (AISC) code for structures such as bridges and buildings, and the American Society of Mechanical Engineers (ASME) codes for pressure vessels and boilers. These codes specify the design stress and include all factors that are considered appropriate for the application by these organizations. Only the strengths and practices of the AISC Code will be used here. Table 10-1 gives the strength specified by the AISC Code for power-driven rivets; for hand-driven rivets the strengths are considerably less.

The reason for the smaller strength in bearing for single shear is that the lines of application of the forces do not coincide, which causes the more complex stress situation illustrated in Fig. 10-5. The AISC Code also specifies 20,000 psi as the tensile and compressive strength of the structural steel members. Greater strengths than those specified in the AISC Code are sometimes used for machinery. There is considerable friction produced between the plates of a riveted joint, which is ignored when designing the joint.

10-4 DESIGN OF RIVETED JOINTS

The practices generally employed in structural design are:

1. Joints are so arranged that the rivets are not placed in tension.
2. The hole is made 0.062 in. larger in diameter than the rivet.

TABLE 10-1 Strength of Rivets

TYPE OF STRESS	STRENGTH, PSI
Shear	15,000
Bearing:	
Single shear	32,000
Double shear	40,000

3. If the hole is punched, it is assumed to be 0.125 in. larger in diameter than the rivet when calculating the tensile stress in the net area. This is to allow for the damage that may be caused in punching.

4. In calculating the shear and bearing stresses, the size of the rivet is used.

5. The edge distance is made equal to twice the diameter of the rivet, in which case the failures shown in Fig. 10-3 will not occur.

6. The minimum pitch (distance between rivets in a row) is 3 times the diameter of the hole.

7. The minimum distance between rows of rivets is twice the diameter of the hole.

8. Where cover plates are used, they are slightly larger than one-half the thickness of the plates to be joined.

The procedure to follow in designing a joint is:

1. Select several rivets whose diameter is approximately equal to the square root of the thickness of the plates to be joined. Standard rivets are available from ⅜ in. in diameter to 1⅝ in. in diameter, in increments of ⅟₁₆ in.

2. Compare the strengths of the rivets in shear and bearing. For rivets in double shear, multiply the shear strength by 2 before comparing. Select the rivet size for which the shear strength and the bearing strength are most nearly equal.

3. Determine the number of rivets to use by dividing the load to be transmitted by the smaller strength of the chosen rivet — shear strength or bearing strength.

4. If the members are in tension, arrange the rivet pattern so that the tension in the net area is greater than the strength of the rivets, keeping in mind the requirements for minimum pitch, edge distance, and distance between rows of rivets.

There are instances, of course, where departing slightly from the best rivet size may be desirable. Such an instance would be to permit an arrangement where the line of action of the load would pass through the center of gravity of the rivet areas. If the joint is long, with a rivet pattern that is repeated, a representative portion may be used rather than the entire joint in performing the calculations, in which case the appropriate portion of the total load must be used.

Example 10-1 In Fig. 10-6 the plates are ⅜ in. thick, the rivets are ⅝ in. in diameter, and the holes are punched. What is the maximum tension load to which the joint should be subjected?

Fig. 10-6 Example 10-1.

Solution The shear strength of the rivet per Table 10-1 is 15,000 psi; thus, the maximum load limited by shear is 3 times the cross-sectional area of the rivet times 15,000; thus

$$0.375 \times 3 \times 15,000 = 16,800 \text{ lb}$$

The bearing strength is 32,000 psi; thus, the maximum load limited by bearing is

$$0.625 \times 0.375 \times 3 \times 32,000 = 22,500 \text{ lb}$$

The tensile strength is 20,000 psi. The holes are punched; therefore, 0.75 is the diameter used in calculating the net area. The net width is

$$7.0 - (3 \times 0.75) = 4.75 \text{ in.}$$

The maximum load limited by tension in the net area is

$$4.75 \times 0.375 \times 20,000 = 35,600 \text{ lb}$$

The 16,800 lb is the smallest load and is, therefore, the maximum load to which the joint should be subjected.

10-5 WELDED JOINTS

Welding is usually cheaper than riveting and has the advantage that an invisible joint may be made. This may be desirable when appearance is a factor. The use of machine frames fabricated by welding, instead of castings, is increasing. As metallic-arc welding is used most frequently, it will be the only kind discussed.

The most common types of joints and loading are shown in Figs. 10-7 and 10-8. A butt weld loaded to put the weld in tension is shown in Fig. 10-7a. A butt weld loaded to put the weld in shear is shown in Fig. 10-7b. A lap weld or double transverse fillet weld which puts the weld in tension is shown in Fig. 10-7c. The lap should be at least 5 times the thickness of the thinner plate. A parallel fillet weld is shown in Fig. 10-7d. This type and loading produce shear in the weld. A single transverse fillet weld is shown in Fig. 10-8a. This type is undesirable, because of the tendency to open up, as shown in Fig. 10-8b, which increases the stress in the weld. A tee joint is shown in Fig. 10-8c, which produces tension in the weld. If one of the welds is omitted, the total length of weld is reduced to one-half, but the stress in the remaining weld is more than doubled, because of the bending that is introduced.

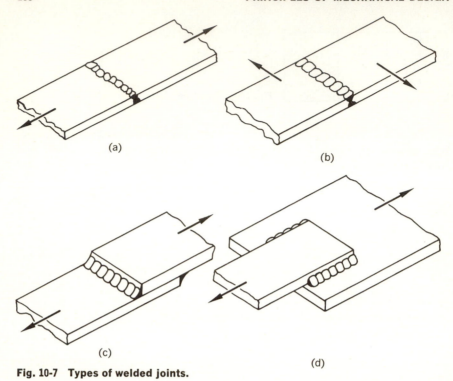

Fig. 10-7 Types of welded joints.

10-6 STRENGTH OF WELDED JOINTS

The strength of a welded joint depends on both the strength of the weld metal and the area of weld metal involved; that is, on the size and length of the weld. For a butt weld that penetrates the entire thickness of the plates, the dimension to use in calculating the strength is the thickness of the plate. For a fillet weld, as shown in Fig. 10-9, the size of the weld is h, but the dimension to use in calculating strength is t, which for the usual weld is equal to $0.707h$.

The size of the weld is largely controlled by the thickness of the mem-

TABLE 10-2 Weld Sizes

THICKNESS OF MEMBER	WELD SIZE
0.10–0.20	0.15
0.20–0.30	0.20
0.30–0.60	0.30
0.60–1.00	0.40
1.00–1.40	0.60

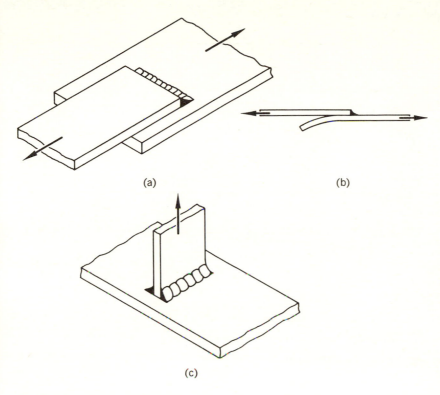

(a)   (b)

(c)

Fig. 10-8   Types of welded joints.

bers being joined. A weld size appropriate for various thicknesses is given
in Table 10-2.

The design stress for several types of joints made by metallic-arc welding in low-carbon steel with coated rods is given in Table 10-3, for both
static and dynamic loads. The strengths for dynamic loads given in Table
10-3 must be divided by a factor to allow for stress concentrations. These
factors depend on the type of joint and are as follows: butt joint = 1.2,
transverse fillet = 1.5, parallel fillet = 2.7, tee joint = 2.0.

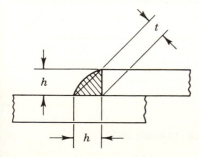

Fig. 10-9   Dimensions of a fillet weld.

TABLE 10-3 Weld Strengths

	DESIGN STRESS, PSI	
WELD TYPE AND LOADING	STATIC LOADS	DYNAMIC LOADS
Butt welds:		
In tension	16,000	8000
In compression	18,000	8000
In shear	10,000	5000
Fillet welds	14,000	5000

The length of weld is found for a butt weld by using the equation

$$L = \frac{P}{hs} \tag{10-5}$$

For a fillet weld the equation is

$$L = \frac{P}{0.707hs} \tag{10-6}$$

where P = load to be transmitted, lb
h = weld size, in.
L = weld length, in.
s = weld design stress, psi

To allow for imperfections caused by starting and stopping a weld bead, 0.25 in. should be added to the calculated length for each separate bead in the joint.

Example 10-2 The joint shown in Fig. 10-10 is subjected to a 3000-lb dynamic load, and the size of the welds is 0.40 in. Find the length of each weld.

Solution The joint is a parallel fillet weld, and the weld is in shear; thus, the strength from Table 10-3 is 5000 psi. This strength must be divided by 2.7, because

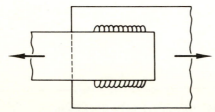

Fig. 10-10 Example 10-2.

the load is dynamic; $5000/2.7 = 1850$ psi. The total length required is found with Eq. (10-6):

$$L = \frac{3000}{0.707 \times 0.40 \times 1850} = 5.75 \text{ in.}$$

The length of each weld is

$$\frac{5.75}{2} + 0.25 = 3.13 \text{ in.}$$

The 0.25 is the allowance for starting and stopping the weld.

10-7 DESIGN OF WELDED JOINTS

In designing a welded joint, a guiding principle is to provide strength by length rather than by size, because it takes a great deal more weld metal to develop the required strength when a large short weld is used than when a small long weld is used. Weld beads with lengths less than 4 times the size of the weld should not be counted as part of the length of the weld for strength calculations. The joint should be arranged to minimize any tendency to pry off the member.

One of the most important principles in the design of welded joints is to arrange the welds so that they are symmetrical in location, size, and length. For members such as the structural steel angle shown in Fig. 10-11, this means that the length of the weld is divided in accordance with the equation

$$L_1 \times a = L_2 \times b \tag{10-7}$$

where a and b are dimensions from the centroidal axis of the section.

In designing a welded assembly, the designer should guard against two tendencies: one is to join many pieces cut from plate; the other is for the design to be influenced by the limitations of the processes used to produce similar parts, such as casting, forging, or riveting. Structural

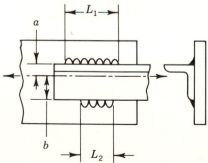

Fig. 10-11 Symmetry applied to welding an angle.

sections and bent plates can often be used to advantage, and it is not poor design to use a steel casting as part of a welded assembly where the cost of the pattern can be justified.

10-8 FURNACE-BRAZED JOINTS

Copper is well suited for the task of joining steel parts, because of the pronounced capillary action that takes place when the parts are heated in close contact. A temperature of about 2000°F is required. Flux is not necessary, and because a controlled-atmosphere furnace is used, oxide or scale does not form on the parts.

Though butt joints have considerable strength, they should be avoided wherever possible. For lap joints, such as those shown in Fig. 10-12, the length of the lap may be found with the equation

$$L = \frac{s_t t}{s_s} \tag{10-8}$$

where L = length of lap, in.

s_t = tensile strength of thinner part, psi

t = thickness of thinner part, in.

s_s = shear strength of brazing metal, psi

When copper is used as the brazing metal, the shear strength is about 22,000 psi. The length of lap found by this equation will produce a strength equal to the parent metal, but includes no factor; thus, the length to use in a design should be at least 1½ times the length calculated.

The clearance between the parts is important. A clearance of 0.002 in./in. of diameter is recommended, though slightly less than or several times this much is satisfactory, which permits a reasonable tolerance for the parts. The parts must be held in the proper relation to each other during

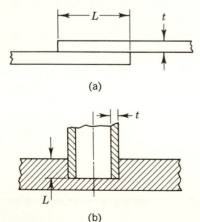

(a)

(b)

Fig. 10-12 Furnace-brazed lap joint.

both heating and cooling. Jigs should be avoided. The parts should be designed to fit together and remain properly assembled at the brazing temperature. The areas to be joined should be nearly vertical when being heated, and a provision should be made for the brazing metal, in the form of a wire, at the upper end of the joint. Whenever air would be trapped, a vent must be provided, because the air would expand and interfere with the flow of the brazing metal.

Other factors that deserve consideration are: brazing should never be done before welding — after welding is acceptable; furnace-brazed parts may be heat-treated without destroying the bond; parts that have been work-hardened lose their strength; case-hardened parts do not braze well, though brazed parts may be case-hardened.

10-9 SOLDERED JOINTS

Soldered joints have relatively little strength — about 5000 psi ultimate shear strength for a short time at room temperature. The creep strength is very low. A joint may lose 40 percent of its strength at a temperature of only 200°F. However, strength is not a prime requirement for some joints — for example, for a joint that must be gastight, or where electrical or heat conductivity is required. Wherever a soldered joint is acceptable it is usually the most economical. Less than 800°F is required to make a soldered joint (for some solders only 300°F); thus the difficulties associated with high temperature are largely avoided.

When a joint must have strength and also be gastight, the strength can be provided by some means that mechanically locks the parts together and the solder used just as a seal. The most desirable thickness of solder in a joint is about 0.004 in. The minimum length of lap should be 4 times the thickness of the thinner part.

QUESTIONS

10-1 What are some of the factors that should be considered in deciding on a joining method?

10-2 What are some of the advantages of riveting?

10-3 Discuss edge distance.

10-4 Why is a punched hole assumed to be 0.125 in. larger than the rivet in calculating the tensile stress in the net area of a riveted joint?

10-5 What are the advantages of welding?

10-6 Why is a long small weld preferred to a short large weld of the same strength?

10-7 What is the function of capillary action in a furnace-brazed joint?

10-8 Discuss welding, heat-treating, and case hardening relative to furnace-brazed parts.

10-9 What are some of the advantages and disadvantages of soldering?

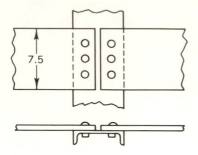

Fig. 10-13   Prob. 10-3.

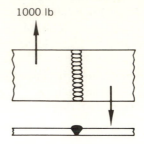

Fig. 10-14   Prob. 10-5.

PROBLEMS

Note: Upon completion of some of these problems, Design Example 1 and Projects 4 and 5 in Chap. 19 are appropriate.

10-1   A double-row lap joint with a total of four 1.0-in.-diameter rivets joins two plates $7/8$ in. thick. It carries a static load of 32,000 lb. What are the shear and bearing stresses?*

10-2   A single-row lap joint, consisting of four $3/8$ rivets, joins two $3/16$-thick plates that are 8 in. wide. The holes are punched. What is the largest tension load to which the joint should be subjected?

10-3   For the joint shown in Fig. 10-13, the rivets are $3/4$, the plates are $1/2$ thick, the channel is 0.579 in. thick, and the holes are drilled. What is the largest tension load to which the joint should be subjected?

10-4   Two plates $5/16$ thick are joined by a butt weld that completely penetrates the plates. The joint is subjected to a dynamic tension load of 1500 lb. How wide should the plates be?*

10-5   For the joint shown in Fig. 10-14, the plates are $7/16$ thick, and the load is static. What is the width of the plates?

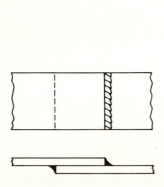

Fig. 10-15   Prob. 10-6.

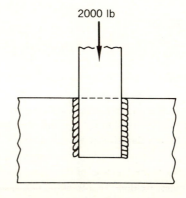

Fig. 10-16   Prob. 10-7.

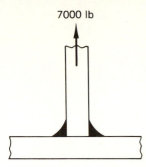

7000 lb

Fig. 10-17   Prob. 10-8.

10-6   For the joint shown in Fig. 10-15, the plates are ⅝ thick and 10 in. wide, and the size of the welds is 0.40. What is the maximum dynamic tension load to which the joint should be subjected?

10-7   For the joints shown in Fig. 10-16, the plates are ⅜ thick, the welds are 0.30, and the load is static. What is the length of each weld?

10-8   For the joint shown in Fig. 10-17, the vertical member is 4 in. wide and is welded for the full width on both sides with a 0.20 weld. If the load is static, what stress is induced in the weld?

10-9   Two tubular members of steel with a tensile strength of 65,000 psi are to be joined by furnace brazing. Both have a ³⁄₁₆ wall and the outside diameter of the inner member is 3 in. What is the recommended length of lap?*

10-10   Two tubular members are to be joined by furnace brazing. They are steel, of 90,000 psi tensile strength. The outer member has a nominal inside diameter of 2 in. and a wall thickness of 0.125. The inner member has a wall thickness of 0.10. Find the length of lap and the upper and lower limit of the two diameters.

SELECTED REFERENCE

Laughner, V. H., and A. D. Hargen: "Handbook of Fastening and Joining Metal Parts," McGraw-Hill Book Company, New York, 1956.

SHAFTS AND COUPLINGS

11-1 INTRODUCTION

Shafts are an essential part of practically every machine. Some of the more important factors unique to shafts that deserve consideration in their design are: adequate torsional strength to transmit the power, sufficient rigidity to prevent interference with the proper performance of bearings and gears, an allowance for the effects of necessary details such as keyways. An important part of any discussion of shafts is the means of connecting shafts to each other or to other machine elements.

11-2 TERMINOLOGY

Shafts involved in power transmission may be either transmission shafts or machine shafts. Transmission shafts are used between the power source and the machine, and machine shafts are a part of the machine itself. Strictly speaking, a shaft is a machine element that rotates and is subjected to torsion, and an axle is a member that is not subjected to torsion and often does not rotate. Thus, the rear axle of most automobiles is really a shaft. A short shaft in a machine is sometimes referred to as a spindle.

11-3 TORQUE AND POWER

Torque is a moment and is, therefore, equal to a force times a distance. In Fig. 11-1, F is the force and r is the distance. T is used to represent torque, and it will be in pound-feet if F is in pounds and r is in feet,

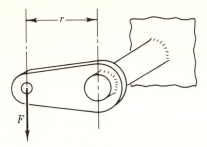

Fig. 11-1 The definition of torque.

but if r is in inches it will be in pound-inches. When dealing with the strength of a shaft, torque must be in pound-inches, because the strength of the material is in pounds per square inch.

Work is defined as the product of a force and the distance through which it moves. Therefore, the work developed by the force F acting at the distance r from the center of rotation in one revolution is $F\,(2\pi r)$. If N is revolutions per minute, then $FN(2\pi r)$ is the number of pound-inches per minute developed. One horsepower is 33,000 lb-ft/min, or 396,000 lb-in./min. Thus,

$$\text{hp} = \frac{FN(2\pi r)}{396,000}$$

but

$$T = Fr$$

and thus

$$\text{hp} = \frac{2\pi TN}{396,000} \quad \text{or} \quad \frac{TN}{63,000} \tag{11-1}$$

11-4 TORSIONAL STRESS

In the study of strength of materials it is shown that when a shaft is subjected to a torque a shearing stress is produced, which varies from zero at the axis of the shaft to a maximum at the outside surface or, as is often said, the extreme fiber. The relation between torque T in pound-inches and the maximum shear stress s_s in psi is

$$s_s = \frac{Tc}{J} \tag{11-2}$$

where J is a property of the cross section of the shaft, known as the polar moment of inertia, and c is the distance from the neutral axis to the extreme fiber.

For a solid circular shaft, c is equal to one-half the diameter and

$$J = \frac{\pi D^4}{32}$$

where D = diameter. Thus, Eq. (11-2) becomes

$$s_s = \frac{TD/2}{\pi D^4/32} = \frac{5.1T}{D^3} \tag{11-3}$$

For a hollow circular shaft

$$J = \frac{\pi(D^4 - d^4)}{32}$$

where d = inside diameter. Thus

$$s_s = \frac{5.1TD}{D^4 - d^4} \tag{11-4}$$

11-5 TORSIONAL DEFLECTION

When a shaft is transmitting torque from one end to the other, there is a tendency for the shaft to twist, as shown in Fig. 11-2. The total angle of twist in degrees for a solid circular shaft of uniform cross section is found with the equation

$$\theta = \frac{584LT}{GD^4} \tag{11-5}$$

For a tubular shaft

$$\theta = \frac{584LT}{G(D^4 - d^4)} \tag{11-6}$$

where L = length of shaft, in.
$\qquad G$ = modulus of rigidity, psi

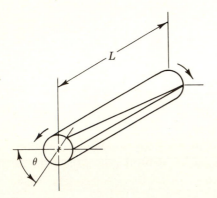

Fig. 11-2 Torsional deflection.

Example 11-1 What are the shear stress and torsional deflection for a 1.75-in.-diameter shaft 15 in. long when transmitting 40 hp at 1200 rpm? The modulus of rigidity is 12,000,000 psi.

Solution Solving Eq. (11-1) for torque,

$$T = \frac{63,000 \text{ hp}}{N}$$

The torque is

$$T = \frac{63,000 \times 40}{1200} = 2100 \text{ lb-in.}$$

The stress is found with Eq. (11-3):

$$s_s = \frac{5.1T}{D^3} = \frac{5.1 \times 2100}{1.75^3} = 2000 \text{ psi}$$

The deflection is found with Eq. (11-5):

$$\theta = \frac{584LT}{GD^4} = \frac{584 \times 15 \times 2100}{12,000,000 \times 1.75^4} = 0.163°$$

11-6 SHAFTS IN BENDING

When a shaft carries only a bending load, it is treated as a beam. For a solid circular shaft subjected to a maximum bending moment M in pound-inches, the maximum stress in psi is found with the equation

$$s_t = \frac{32M}{\pi D^3} \tag{11-7}$$

For a tubular shaft

$$s_t = \frac{32M}{\pi D^3} \times \frac{1}{1 - C^4} \tag{11-8}$$

where $C = d/D$.

If the shaft is of uniform cross section, the deflection may be found by using the appropriate beam-deflection equation for the type of loading. If the shaft is not of uniform cross section, as is often the case, the problem is more complex.

If the shaft is short and is subject to a considerable bending load, it should be checked for transverse shear in the same manner as a short beam is checked.

11-7 SHAFTS WITH COMBINED TORSION AND BENDING

Shafts involved in the transmission of power by means of belts, gears, and chains are subjected not only to torsion but to bending. Combined loading was discussed in Sec. 7-8. It will be remembered that in order to calculate the effect of the combination of loads a theory of failure must be chosen, also that for ductile metals the maximum-shear-stress theory is used, and for brittle metals the maximum-normal-stress theory is used.

According to the maximum-shear-stress theory the maximum shear stress due to the combined load is found with Eq. (7-6), which is

$$s_{s,max} = \sqrt{s_s^2 + \frac{s_t^2}{4}}$$

where s_s = shear stress found with Eqs. (11-3) or (11-4)
      s_t = tensile stress found with Eqs. (11-7) or (11-8)
Thus, for a solid circular shaft, Eq. (7-6) becomes

$$s_{s,max} = \frac{5.1}{D^3} \sqrt{T^2 + M^2} \tag{11-9}$$

where T = torque, lb-in.
      M = bending moment, lb-in.
And for a hollow circular shaft

$$s_{s,max} = \frac{5.1}{D^3} \sqrt{T^2 + M^2} \times \frac{1}{1 - C^4} \tag{11-10}$$

According to the maximum-normal-stress theory the maximum tensile stress due to the combined load is found with Eq. (7-5), which is

$$s_{t,max} = \frac{s_t}{2} + \sqrt{s_s^2 + \frac{s_t^2}{4}}$$

which for a solid circular shaft becomes

$$s_{t,max} = \frac{5.1}{D^3} (M + \sqrt{T^2 + M^2}) \tag{11-11}$$

and for a hollow circular shaft

$$s_{t,max} = \frac{5.1}{D^3} (M + \sqrt{T^2 + M^2}) \times \frac{1}{1 - C^4} \tag{11-12}$$

The term $\sqrt{T^2 + M^2}$ is often referred to as the equivalent twisting moment, and is defined as the fictitious torsional moment that will induce the same shear stress in the shaft as the actual torsion and actual bending combined.

11-8 SHOCK FACTORS

The equations for combined loading, Eqs. (11-9) to (11-12) are valid for stationary shafts when the loads are steady or gradually applied, and for rotating shafts when the torque load is steady or gradually applied. For other conditions, a factor must be included. However, one of the loads may be gradually applied and the other suddenly applied. Thus, the same factor does not apply to both, and the usual method of finding the size with the working load and design stress cannot be used. A way of dealing with this situation is to remove the strength factor b from Eq. (7-1) and apply a factor individually to the moments M and T to compensate for its removal. The factor applied to M is identified as C_M, and that for T as C_T. The corrected moment $M' = M \times C_M$, and the corrected moment $T' = T \times C_T$. The corrected moments are used the same as the moment M and T in the equations for combined loading.

The values for C_M and C_T depend on whether the shaft is stationary or rotating, and on the manner in which the load is applied. For stationary shafts and a suddenly applied load, both C_M and C_T are between 1.5 and 2.0. For a rotating shaft with a steady bending load, C_M is 1.5; because of the rotation, the bending load is constantly being rather suddenly applied. For a rotating shaft with minor shock loads, C_M is between 1.5 and 2.0, and C_T is between 1.0 and 1.5. For heavier shock loads, these factors are considerably greater, the amount depending on the magnitude of the shock. The factors C_M and C_T only compensate for the removal of b from Eq. (7-1), and except for this the procedure for finding the size of a part is unchanged.

Example 11-2 A rotating shaft, 2 ft long and 1.25 in. in diameter, supports a 600-lb load in the center and transmits a torque of 2500 lb-in. applied as a light shock load. What is the maximum shear stress due to the combined load?

Solution The bending moment is found with the beam equation

$$M = \frac{PL}{4} = \frac{600 \times 24}{4} = 3600 \text{ lb-in.}$$

The bending is steady. Therefore, $C_M = 1.5$. The torque is a light shock load. Thus, $C_T = 1.5$. The stress is found with Eq. (11-9):

$$s_{s,\text{max}} = \frac{5.1}{D^3} \sqrt{(C_T T)^2 + (C_M M)^2}$$

$$= \frac{5.1}{1.25^3} \sqrt{(1.5 \times 2500)^2 + (1.5 \times 3600)^2}$$

$$s_{s,\text{max}} = 17,200 \text{ psi}$$

11-9 STRESS CONCENTRATIONS

In order to best serve their purpose, most shafts must have such items as shoulders and keyways, which may cause a stress concentration. If they cannot be eliminated, their effect should be reduced to a minimum and an appropriate allowance made for them in designing the shaft. Stress

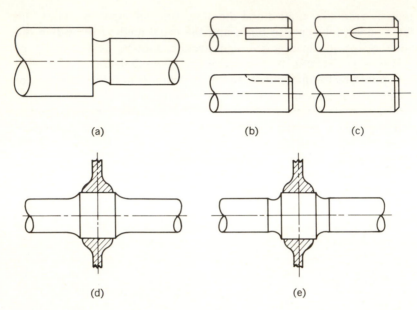

(a)   (b)   (c)

(d)   (e)

Fig. 11-3   Stress concentrations in shafts.

concentrations generally occur wherever there is an abrupt change in cross section.

When the diameter of a shaft is reduced to provide a shoulder, the magnitude of the stress concentration increases with an increase in the difference in diameters. The effect of changes in diameter is reduced by increasing the fillet radius. When the difference in diameters is small, a plain fillet radius of reasonable size will not provide a definite shoulder. The form shown in Fig. 11-3a is often used in this case.

Keyways cause a stress concentration, the magnitude of which depends on the number and type. The type shown in Fig. 11-3b is preferred to that in Fig. 11-3c. The strength of the shaft should be reduced from 15 to 25 percent to allow for the effect of keyways. If the keyway is at a position on the shaft which is not highly stressed, an allowance is not necessary.

When a part with a rather long hub is press-fitted to a shaft, a stress concentration of considerable magnitude may be created. Two methods of reducing this stress concentration are shown in Fig. 11-3d and e.

11-10   CRITICAL SPEED

If the center of gravity of a shaft, or an assembly such as a shaft and gear, lies on the axis of rotation, it will run smoothly at any speed. Because of imperfections and deflections that always exist, the center of gravity is not exactly on the axis of rotation. In such a case, a centrifugal force is produced that tends to increase the deflection of the shaft. The strength in the shaft resists this, but as the rotational speed increases, a speed is reached where a violent vibration takes place. This speed is known as the critical speed. If the speed continues to increase, the vibration decreases as the assembly begins rotating about an axis other than its original axis of rotation.

The vibration that takes place at the critical speed can damage or even destroy a machine. A well-constructed machine can be successfully operated close to or above the critical speed; however, as a general rule, the maximum operating speed should not exceed 80 percent of the critical speed.

If a shaft has a concentrated load located somewhere along its span, the critical speed may be found using the equation

$$N_c = \frac{188}{\sqrt{Y}} \tag{11-13}$$

where N_c = critical speed, rpm
Y = deflection of the shaft, in.
The deflection is found using the methods for beams.

If the shaft is steel, has a solid circular cross section, and is supported by thin bearings or self-aligning bearings, Eq. (11-13) becomes

$$N_c = 387{,}000 \frac{D^2}{ab} \sqrt{\frac{L}{P}} \tag{11-14}$$

where D = diameter, in.
L = distance between supporting bearings, in.
P = load, lb
a and b = distance from load to bearings, in.
If the shaft is rigidly supported in long bearings, the equation is

$$N_c = 387{,}000 \frac{D^2 L}{ab} \sqrt{\frac{L}{Pab}} \tag{11-15}$$

For these equations, the shaft must be of almost uniform diameter — small shoulders and reliefs may be ignored. If the weight of the shaft is relatively small, it is usually ignored, but if it is to be included, one-half the weight is added to the load.

Example 11-3 Assume that the shaft of Example 11-2 is mounted on narrow bearings, and find the critical speed.

Solution The critical speed is found with Eq. (11-14):

$$N_c = 387{,}000 \frac{D^2}{ab} \sqrt{\frac{L}{P}}$$

$$= 387{,}000 \frac{1.25^2}{12 \times 12} \sqrt{\frac{24}{600}}$$

$$= 836 \text{ rpm}$$

11-11 SHAFT DESIGN

A shaft may be designed on the basis of either strength or rigidity. If strength determines the design of a shaft, the type of load (that is, whether it is steady, fluctuating, reversing, or a shock load) and stress concentrations (such as keyways and shoulders) are to be allowed for. If rigidity or deflections are to determine the design, both transverse and torsional deflections are limited. The factors pertaining to both strength and rigidity must be considered in the design, though only one actually dictates the size and details that constitute the design.

The form of the cross section has a great effect on the design of a shaft. Only circular cross sections will be considered here, because they are by far the most common. As the center of a solid shaft contributes little in most situations, a tube may be the best choice, even allowing for its greater cost, especially when weight is a factor.

For a shaft designed on the basis of strength, the material and any treatment such as heat-treating has considerable effect. Cold-finished low-carbon steel should be considered first. Cold-drawn steel is often used, because the strength is greater and it also has a smooth surface. However, it also has higher residual stresses that may cause objectionable distortion when the shaft is machined. Heat-treating increases the strength and also the hardness, which may be important where wear is involved, but the effect of stress concentrations is increased, because of the greater notch sensitivity.

The deflection of a shaft when used in a machine is often more important than its strength, and if it is satisfactory with regard to deflection, it is generally strong enough. The deflection of a shaft is not dependent on the strength of the metal that it is made from; therefore, if deflection dictates the design, the more costly medium-carbon or alloy steel should not be used.

The deflection due to bending for shafts carrying a gear should not exceed, at the gear, 0.005 in./in. of gear face. For shafts supported by plain bearings, the deflection should not exceed 0.0015 times the distance from the point of application of the load to the center of the bearing in inches. The deflection due to torsion should be limited to less than 0.1°/ft for the usual situation, though as little as one-half this value may be more appropriate for long shafts or for those subjected to suddenly reversed loads. For transmission shafts (from power source to machine), the deflections that are acceptable are many times as great as given above.

In deciding upon a diameter, consideration should be given to standard sizes, even if the shaft is to be machined all over, but especially if little or no machining is to be performed.

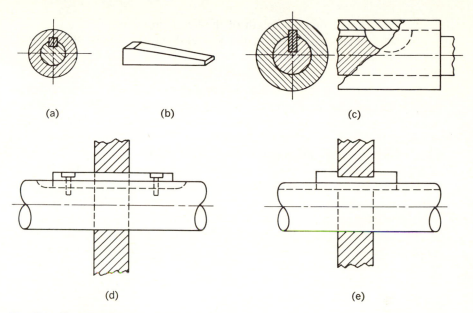

Fig. 11-4 Types of keys.

11-12 TYPES OF KEYS

Keys are used to prevent relative rotation between a shaft and a part, such as a gear or pulley, that is mounted on the shaft. To say it another way, a key is used to transmit a torque from a shaft to a hub, or from a hub to a shaft. Several types that are frequently used are shown in Fig. 11-4. The square key, shown in Fig. 11-4a, is the most common. The taper key, shown in Fig. 11-4b, is driven into place and tends to remain in place, but this may cause the hub to tip or to be made eccentric with the shaft. The Woodruff key, shown in Fig. 11-4c, has the advantage, when used with a tapered shaft, that it will rock in the key seat to align itself with the keyway in the hub, which permits greater tolerance in machining. However, it requires the removal of considerable metal from the shaft. The purpose of the feather key, shown in Fig. 11-4d and e, is to transmit torque and still permit axial motion of the hub. If the axial motion is required while torque is being transmitted, two feather keys should be used, placed on opposite sides of the shaft, because only about one-half of the force is required to produce the axial motion when two keys are used as when a single key is used. The square key is used most often. It will be discussed in detail here.

11-13 KEY DESIGN

The forces acting on keys are rather complicated, and two simplifying assumptions are made in the design of a keyed connection:

1. The force along the length of the key is uniform.
2. The force on the side of the key and on the side of the keyway acts at the radius of the shaft.

Referring to Fig. 11-5, in transmitting the torque there is a tendency to shear the key, which is resisted by the force that is equal to the shear strength times the area in shear. There is a tendency for the hub to crush the key, which is resisted by the compression strength of the key times the area found by multiplying one-half the height by the length. There is also the tendency of the key to crush the shaft, which is resisted by the compression strength of the shaft times the area equal to one-half the height times the length.

Torque is a moment; therefore, it must be resisted by a moment. The strength of the key in psi times the area is a force, which multiplied by the radius of the shaft is a moment. Stating all of this mathematically,

$$T = W \times L \times s_s \times \frac{D}{2}$$

Thus

$$s_s = \frac{2T}{WLD} \tag{11-16}$$

Also

$$T = \frac{h}{2} \times L \times s_c \times \frac{D}{2}$$

and thus

$$s_c = \frac{4T}{hLD} \tag{11-17}$$

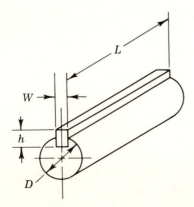

Fig. 11-5 Key dimensions.

where T = torque, lb-in.
     L = length of key, in.
    W = width of key, in.
    h = height of key, in.
    D = diameter of shaft, in.
    s_s = shear stress, psi
    s_c = compression stress, psi

If the shaft has greater compression strength than the hub, a greater portion of the height of the key could be used in the hub. Equation (11-17) would have to be modified to allow for this. The size of the key should be limited to about one-fourth the shaft diameter.

Frequently, the key is designed so that it will fail before either the shaft or the hub, because it is easier and less costly to replace the key. Plain medium-carbon steel is often used for keys. The key should fit snugly in the keyway of both the shaft and the hub, especially at the width. This is to prevent failure due to the key rolling in the keyways. A setscrew is sometimes used to force the key against the bottom of the keyway and thus prevent it from working out.

Example 11-4 What shear stress and what compression stress is induced in a key 0.50 in. square and 3 in. long, placed in a 2.0-in.-diameter shaft, if 200 hp at 2000 rpm is transmitted?

Solution The torque is

$$T = \frac{63{,}000 \times 200}{2000} = 6300 \text{ lb-in.}$$

The shear stress is found with Eq. (11-16):

$$s_s = \frac{2T}{WLD} = \frac{2 \times 6300}{0.5 \times 3 \times 2} = 4200 \text{ psi}$$

The compression stress is found with Eq. (11-17):

$$s_c = \frac{4T}{hLD} = \frac{4 \times 6300}{0.5 \times 3 \times 2} = 8400 \text{ psi}$$

11-14 SPLINES

A spline may be thought of as a series of keys integral with the shaft, the hub having a series of keyways to match. A spline permits a much more compact connection between a hub and shaft. Splines are used where the hub is fixed relative to the shaft, also where it must slide along the shaft.

There are several systems of splines: the parallel-side spline is similar to the key and keyway idea; the involute spline has teeth similar to the teeth

Fig. 11-6 An involute spline.

of a gear. The tooth form for an involute spline is shown in Fig. 11-6. Two of the reasons why the involute spline is preferred over the parallel-side spline are greater strength and more economical manufacture.

There are several types of fits that vary regarding where on the tooth form the mating parts are in contact. There are also several classes of fits that vary in the clearance between the mating parts.

Serrations are similar to splines, except that the tooth form is more shallow. They are used principally where axial motion is not required. A relatively large number of teeth is often used, which permits locating the hub closer to a given radial position.

11-15 COUPLINGS

A common problem in mechanical design is that of fastening two shafts together. Some of the reasons for joining shafts with a coupling are:

1. To join units that are manufactured in different locations or are more convenient to handle as smaller units
2. To join standard units to accomplish a special purpose
3. To provide for misalignment of the shafts
4. To prevent transmission of an overload
5. To reduce the transmission of shock or vibration
6. To rapidly connect or disconnect the shafts as required by the operation of the machine
7. To connect a shaft that is rotating to one that is not
8. To allow for axial motion of the connected shafts caused by thermal expansion

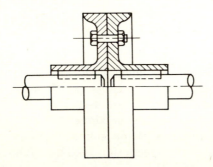

Fig. 11-7 A rigid coupling.

As there are a number of reasons for joining shafts, there are a number of different kinds of couplings in order to satisfy the various requirements.

11-16   RIGID COUPLINGS

A rigid coupling does not provide for any misalignment of the shafts, nor does it tend to reduce the transmission of shock or vibration. The most common type of rigid coupling, known as the flanged shaft coupling, is shown in Fig. 11-7.

The torque in the left shaft is transmitted to the hub, causing compression stresses on the shaft, key, and hub, and causing shear in the key. As the load passes from the hub to the flange, a shear stress is induced in the flange. The bolts carry the load from the left flange to the right flange, producing a bearing stress in the flanges and the bolts, and a shear stress in the bolts at the face of the flanges. The load passes on to the right shaft, producing the same stresses in the right members as have been described.

The method of calculating the stresses in keyed and bolted connections has been covered, which leaves only the shear in the flange at the hub to be considered. The force at this point is equal to the torque divided by the radius of the hub, and the area involved is the circumference of the hub times the thickness of the flange. Thus, the shear stress is

$$s_s = \left(\frac{T}{D/2}\right) \div (\pi D t) = \frac{2T}{\pi D^2 t} \tag{11-18}$$

where D = hub diameter, in.

$\quad\;\; t$ = flange thickness, in.

The diameter of the hub is generally arrived at by the application of an empirical rule, which is: the diameter of the hub is equal to 1.75 times the shaft diameter plus 0.25 in.

The rim at the outside diameter is to reduce the danger of the bolt heads or nuts catching a workman's clothing. The load was assumed to be divided equally among the bolts. For this assumption to be valid, each bolt must snugly fit the hole. This requires that the holes in both flanges be drilled and reamed while assembled as well as the use of bolts whose diameters are held to a close tolerance.

11-17   FLEXIBLE COUPLINGS

If two shafts joined by a rigid coupling are not in perfect alignment, stresses due to bending are induced that are reversed with each revolution of the shafts. These stresses greatly reduce the life of the shafts. This misalignment also causes an additional load on the bearings that support the

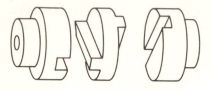 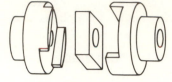

Fig. 11-8   An Oldham coupling.   **Fig. 11-9   An American flexible coupling.**

shafts. Thus, whenever two units, such as a motor and a pump, are to have their shafts joined, a flexible coupling should be used.

Some couplings derive their flexibility from their kinematic arrangement. An example of this type is the Oldham coupling, shown in Fig. 11-8. It is well suited to shafts whose center lines are parallel but do not coincide. The center part has tongues at right angles to each other, which fit in slots in the end members. The tongues are free to slide in their slots, and these motions accommodate the misalignment.

Another coupling, similar in principle to the Oldham, is the American flexible coupling, shown in Fig. 11-9, which will accommodate misalignment of parallel shafts and some angular misalignment. Both of these couplings may be used to compensate for slight axial motion.

Some couplings derive their flexibility from the resilience of one or

Fig. 11-10   A flexible coupling. (*Dodge Manufacturing Corp.*)

Fig. 11-11 A flexible coupling. *Ajax Flexible Couplings Co., Inc.)*

more of their parts. The simplest coupling of this type is a piece of rubber tubing clamped to the ends of the two shafts. It is satisfactory for all types of misalignment, but only for the transmission of small amounts of power. The coupling shown in Fig. 11-10 is based on the principle of the rubber tubing mentioned above, and is satisfactory for all types of misalignment; it is also capable of transmitting considerable power. The coupling shown in Fig. 11-11 accommodates a small amount of misalignment of all types and relatively large axial motion between the shafts. The steel pins ride in self-lubricating bronze bushings which are bonded to the rubber bushings. Couplings of this type are available with a capacity of 15,000 hp. All-metal couplings are available that allow for various amounts of misalignment, some with capacities of over 100,000 hp.

11-18 UNIVERSAL JOINTS

When the angular misalignment between two shafts is considerable, flexible couplings cannot be used, and some form of universal joint is required. A universal joint is a linkage that permits rather large angular motions in all directions. The most common type, known as Hook's coupling, consists of three major parts: two forks, or yokes, and a center piece called a block or spider, as shown in Fig. 11-12.

The angular displacement of the driven shaft does not equal the

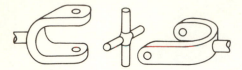

Fig. 11-12   A universal joint.

angular displacement of the driving shaft throughout a revolution of the shaft. Each shaft makes the same total number of revolutions in a given time, but the angular velocity varies, fluctuating through two maximums and two minimums every 360° of rotation. This cyclic variation results in a torsional vibration. The magnitude of the variation in angular velocity increases with an increase in the acute angle between the shafts. Thus, this angle should be less than 15°, though for very low speeds, such as hand operation, 45° may be used.

If two universal joints are used, their yokes can be so oriented that the variations in angular velocity cancel out. If all three shafts are in the same plane, and the angle between the driving shaft and the intermediate shaft is equal to the angle between the driven shaft and the intermediate shaft, the proper orientation of the yokes is shown in Fig. 11-13.

If, for the arrangement shown in Fig. 11-13, one of the yokes on the intermediate shaft is splined, the yoke can move axially along the intermediate shaft. This permits either the driving shaft or the driven shaft to move up and down while transmitting power. This is the arrangement used on automobiles for connecting the transmission and the differential.

There are single universal joints that involve a rather complex arrangement of balls and races that transmit power at a constant angular velocity. They are used in vehicles where the wheel is used for both driving and steering.

Another method of connecting the driving shaft to a driven shaft that must change location is to use a flexible shaft. Where the power to be transmitted is not large, this is an excellent solution. These units are designed to transmit power in a given direction of rotation, and when used

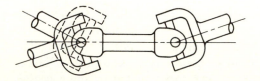

Fig. 11-13   The proper orientation of yokes.

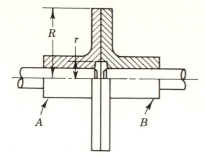

Fig. 11-14 A slip coupling.

with the direction of rotation reversed the capacity is considerably reduced. Also, the radius of curvature to which the shaft is bent has a great effect on the power-transmitting capacity.

11-19 OVERLOAD RELEASE COUPLINGS

It is often desirable to limit the torque that is transmitted to a pre-determined value, in order to prevent damage to the machine or to the material in process when an overload occurs. The overload may be some-thing that is expected to happen frequently, or it may be due to the un-expected failure of something to perform in the prescribed manner. A coupling that would be satisfactory for the unexpected overload probably would not be acceptable for the frequent overload.

The basic arrangement of one form of overload release coupling, known as a slip coupling, is shown in Fig. 11-14. Units of this type will be used as design projects, and in order to encourage independent solutions greater detail is not included in the illustrations. Part A is securely attached to the left shaft; part B is free to move to left and right. Constant force on B presses it against A, and the friction between the two transmits the torque. The magnitude of the torque can be controlled by adjusting the force on B. The torque transmitted by this type of coupling is

$$T = fP \frac{R + r}{2} \tag{11-19}$$

where T = torque, lb-in.
    f = coefficient of friction
    P = axial force, lb
    R = outside radius of the friction faces, in.
    r = inside radius of the friction faces, in.
The coefficient of friction will be discussed in Chap. 16, but a value of 0.1 may be assumed for work at this time.

This type of coupling has the advantage that it is not necessary to reset

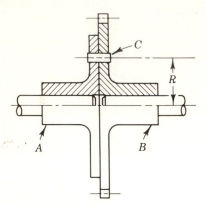

Fig. 11-15 A shear-pin type of overload release coupling.

it after overload, and the disadvantage that if the overload goes unnoticed the coupling may be destroyed by the great amount of heat generated.

The basic arrangement of a shear-pin type of overload release coupling is shown in Fig. 11-15. Parts A and B are attached to the shafts; without the pin C, part B is free to rotate relative to A. B could be a gear or sprocket. The torque in pound-inches that is transmitted is equal to the load in pounds at which the pin shears times the distance R in inches.

The advantage of this type is that it will not be destroyed if the overload goes unnoticed, and the disadvantage is that replacing the pin takes time.

There are many other methods of providing for release when overload occurs, some of which use a steel ball held in an indentation by a spring; when the overload takes place, the ball is forced out of the indentation against the spring. The torque at which the ball is forced out can be varied by an adjusting screw that changes the load produced by the spring. The indentation for the ball could be placed in a shaft or a flange, and more than one ball could be used. This type is, of course, not suitable for large torques.

QUESTIONS

11-1 What are some of the factors that should be considered in the design of shafts?

11-2 What stress is produced in a shaft subjected to torque?

11-3 Discuss stress concentrations as applied to shafts.

11-4 What does the term critical speed mean?

11-5 What is the purpose of a key?

11-6 What is a feather key?

11-7 Why is it important that a key snugly fit the keyway of both the shaft and the hub at the width?

11-8 What is a spline?

11-9 Give four reasons for joining shafts with a coupling.

11-10 Why are two universal joints often used when the angle between the shafts is relatively small?

11-11 What are the two basic types of overload release couplings?

PROBLEMS

Note: After solving some of these problems, study of Design Example 2 and completion of Project 6 or 7 in Chap. 19 is *recommended.*

11-1 What horsepower does 1000 lb-ft at 600 rpm represent?*

11-2 What shear stress is developed in transmitting 10 hp at 800 rpm with a 1-in.-diameter shaft?*

11-3 In Prob. 11-2, what is the stress if a 1-in. tube with a 0.125-in. wall is used?

11-4 If the shaft is steel and 1 ft long, what is the deflection in Probs. 11-2 and 11-3?

11-5 If a 2-in.-diameter steel shaft is stationary and is subjected to a steady bending moment of 300 lb-in. and a gradually applied torque of 400 lb-in., what is the maximum shear stress and the maximum tensile stress due to the combined loading?*

11-6 What is the critical speed of a 1½-in. shaft, 1.5 ft long, that has a 50-lb load in the center and is supported by self-aligning bearings?

11-7 What is the critical speed of the shaft in Prob. 11-6 if the shaft is rigidly supported in long bearings?

11-8 What shear and compression stress is developed in a key 0.25 by 0.25 in., 1.5 in. long, in a 1¼-in. shaft, when transmitting 9 hp at 900 rpm?

11-9 A slip coupling has friction faces with a large diameter of 8 in., a small diameter of 3 in., and an axial force of 50 lb. What torque will it transmit?*

11-10 For a slip coupling, what axial force is required to transmit 5 hp at 1500 rpm, if the outside diameter of the friction faces is 6 in. and the inside diameter of the friction faces is 2 in.?

SELECTED REFERENCE

"Power Transmission Handbook & Directory," *Power Transmission Design Magazine,* Cleveland, 1964.

12

PLAIN BEARINGS

12-1 INTRODUCTION

When one mechanical member rests on another, and there is relative motion between them, they constitute a bearing. However, bearings are generally regarded as separate elements which permit controlled relative motion between members forced together, and because of their design and material they achieve a much more satisfactory performance than would be produced by allowing the members to come in contact.

There are two basic kinds of bearings: those in which the parts in contact slide, and those in which there is rolling between the parts in contact. The rolling bearings will be discussed in the next chapter.

When there is relative motion between surfaces in contact, friction is produced and power is absorbed in overcoming it. Lubrication is used to reduce friction and also to reduce wear. Therefore, friction, wear, and lubrication will be discussed in this chapter.

12-2 TYPES OF PLAIN BEARINGS

Plain bearings primarily intended to resist a load that is applied perpendicular to the axis of the shaft are radial bearings, and those intended to resist an axial load are thrust bearings. Radial bearings are also called journal bearings. A journal is that portion of a shaft which is in contact with the bearing. A full bearing completely surrounds the journal, and a partial bearing does not. A thrust bearing may be the type that is placed at the end of the shaft, or it may be the type that rests against a collar which is attached to, or is integral with, the shaft. Another kind of bearing consists of two flat parallel surfaces, between which there is linear motion. This type is referred to as a slider bearing.

12-3   FRICTION

Friction in some instances is desirable, and in others undesirable. Friction is essential to the operation of clutches and brakes, but in the case of bearings an effort is made to reduce it, because of the power it consumes or the heat that is generated. Another reason for reducing friction is to reduce the wear that usually accompanies a relatively large amount of friction.

Friction may be initially classed as either sliding or rolling. Friction may also be classified as static, which must be overcome to start sliding, or kinetic, which must be overcome to maintain sliding. Though it is kinetic friction which must be overcome in operation, static friction often determines the capacity of the driving unit when motion must be started under a full load.

Sliding friction may take place between surfaces either with or without lubrication. When there is no intentional lubrication and little load, and the surfaces are of average roughness, the situation, greatly enlarged, is as shown in Fig. 12-1. When a load is applied, there is too little area in contact to support it, and peaks of the surfaces are flattened until sufficient area is produced. If the surface is extremely rough, a great deal of deformation must take place. With surfaces of average roughness and a moderate load, sliding between the surfaces will cause a tendency for the peaks and valleys to interlock. Force is required to accomplish the deformation necessary to permit motion. This force is friction.

When two surfaces that are absolutely clean are brought in contact, the high spots weld together, and to produce motion these welds must be broken. Welding then takes place at other spots. Thus, the friction is many times as great as that for surfaces that are, for all practical purposes, clean and dry.

A film will quickly form on any surface that is produced, because of the oxygen and moisture in the air, if from nothing else. These absorbed films, which are extremely thin and very tough, serve to keep the bare metals from coming in contact. However, when such surfaces are in sliding

Fig. 12-1   Two surfaces in contact.

contact, these films may be momentarily swept away from small areas, exposing bare metal, and the welding phenomenon described above will take place. Thus, it is evident that friction will depend on the materials in contact and the roughness of the surfaces.

In another form of sliding friction, the surfaces are separated by a layer of lubricant at all times. In this case, the friction will depend only on the lubricant. This will be discussed in Sec. 12-5. If a cylindrical or spherical roller is made to roll along a flat surface under a load that presses it against the surface, both the roller and the surface will deform. This deformation will change the geometry so that a very slight sliding may occur. All this adds up to what is referred to as rolling friction. The most important fact concerning rolling friction as applied to bearings is that it is a great deal less than sliding friction for comparable conditions.

12-4 WEAR

In Sec. 3-11, wear was defined as the unintentional displacement or removal of particles from a surface. When discussing friction, reference was made to Fig. 12-1 and it was said that the peaks would be flattened until there was sufficient area to support the load. If the deformation was great enough to exceed the elastic limit of the metal, the peaks would be permanently flattened, and for all practical purposes the effect is the same as a loss of metal. Thus, for a journal bearing with a rough surface the clearance could be increased greatly in a short time, and though there is no loss of metal this must be regarded as wear.

Again referring to Fig. 12-1 and the explanation of friction as involving the tendency of the peaks and valleys to interlock, if the surface were sufficiently rough, the interlocking would be of such a magnitude that sliding would not cause the slight deformation as was the case with friction, but would cause parts of the surfaces to be torn off. This is wear. This action will also increase the friction.

If the deformation, and even the tearing off of minute particles, is not too pronounced, it may be regarded as "wearing in." The immediate performance of the bearing will be improved. However, such a bearing will not have the combination of load-carrying capacity and life of a bearing where, because of its smoother surface, this action did not take place. Producing the smoother surfaces would be more expensive, and if the bearing can be made large enough to compensate for the reduced capacity, or if the shorter life is acceptable, designing for wearing in is not a poor practice.

If the surfaces of a bearing were separated at all times by a layer of lubricant, there would be practically no wear.

12-5 FLUID LUBRICATION

A lubricant is any substance that will form a film between surfaces, thus preventing them from actually coming in contact. The major reasons for using a lubricant are to reduce friction, to reduce wear, and to carry away the heat generated in the bearing. A lubricant is classified as fluid, grease, or solid. Grease and solid lubricants will be discussed in Sec. 12-8. The most common fluid lubricant is oil, though many other fluids are used in special situations, including water, air, and gasoline.

The most important property of an oil with respect to lubrication is viscosity, which is a measure of its tendency to flow. As the temperature of an oil is increased, the viscosity decreases, and it flows more readily. A high viscosity can support a greater load, but it also has greater internal friction. When the journal speed is high, the internal friction of the oil is greater. Thus, the viscosity must be carefully chosen for a particular application, keeping in mind that it is the viscosity at the temperature of the oil for normal operation that must be considered.

Discussions of viscosity are confusing, because of the number of different ways in which it is expressed. Absolute viscosity is a mathematical concept. The unit in the metric system is the poise or centipoise, where 100 centipoises is equivalent to 1 poise. In the English system, the unit of absolute viscosity is the reyn. Kinematic viscosity is absolute viscosity divided by density; the unit in the metric system is the stoke or centistoke, 100 centistokes being equivalent to 1 stoke. There are also a number of arbitrary units for expressing viscosity that have been developed to facilitate its measurement. The most common is the Saybolt universal second (SUS), which is the time in seconds required for a quantity of oil to run out of the cup of a Saybolt viscometer.

Stokes can be converted to poises by multiplying by the density of the oil in grams per cubic centimeter. The density of most lubricating oils is approximately 0.9 grams/cu cm. The approximate relation between kinematic viscosity and Saybolt viscosity units is

$$\nu = 0.22S - \frac{180}{S} \tag{12-1}$$

where ν = kinematic viscosity, centistokes
$\quad S$ = SUS
In converting from one unit to another, the temperatures are equal; that is, if the viscosity in centistokes at 100°F is converted to poises, the viscosity in poises will be at 100°F. The typical variation of viscosity with temperature for several oils is given in Fig. 12-2.

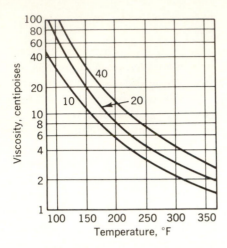

Fig. 12-2 Variation of viscosity with temperature.

Example 12-1 What is the viscosity in centipoises for an oil that has a viscosity of 200 SUS?

Solution The viscosity in centistokes is found by using Eq. (12-1):

$$\nu = 0.22 \times 200 - \frac{180}{200} = 43.1 \text{ centistokes}$$

Centipoises is equal to centistokes times density, which is approximately 0.9. Thus

$$43.1 \times 0.9 = 38.8 \text{ centipoises}$$

12-6 TYPES OF BEARING OPERATION

In the discussion of bearings, the quantity ZN/p is of special interest. Z is the absolute viscosity in centipoises, N is the speed of the journal in revolutions per minute, and p is the bearing pressure in pounds per square inch of projected area. The projected area is the diameter of the journal

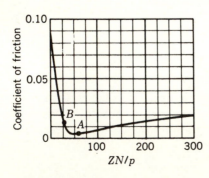

Fig. 12-3 The three realms of bearing operation.

times the bearing length. Three important factors affecting the performance of a bearing are present in this expression. If another important factor, the coefficient of friction, is plotted against ZN/p, a curve similar to that shown in Fig. 12-3 is obtained. This curve illustrates the three realms of bearing operation, which are:

1. Perfect lubrication, also called thick-film or full-film lubrication, which occurs to the right of point A. In this type, the surfaces in the bearing are separated by a relatively thick continuous film of oil, which prevents the surfaces from coming in contact.

2. Boundary lubrication, which takes place to the left of point B, is the situation where bare metal-to-metal contact of the surfaces is prevented only by the thin absorbed films.

3. Imperfect lubrication, also called thin-film or mixed-film lubrication. This type is a combination of perfect and boundary lubrication; that is, the surfaces are, to an extent, separated by a film of lubricant. However, this film is relatively thin and is not continuous; therefore, there are spots where the surfaces are in contact. This type occurs between points A and B.

12-7 PERFECT LUBRICATION

The type of operation known as perfect lubrication can be accomplished by forcing oil into the bearing under pressure, which requires a pump and plumbing. It is called hydrostatic lubrication. The pump can be started before motion in the bearing is permitted, which ensures that there will be a fluid film between the surfaces. Thus, the material of the surfaces is not

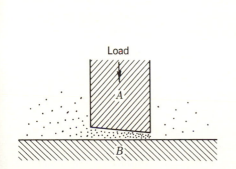

Fig. 12-4 Hydrodynamic lubrication.

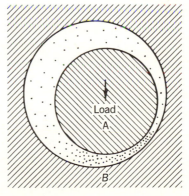

Fig. 12-5 Hydrodynamic lubrication in a journal bearing.

a factor in the performance of the bearing. This degree of complexity is justified in the case of large bearings, such as power generators, or those supporting the propeller shaft of a ship.

Another way of providing perfect lubrication is to design the bearing in a manner such that it will generate the oil pressure needed to provide the thick film. This is called hydrodynamic lubrication. Figure 12-4 shows two surfaces, A and B, separated by a film of oil. Surface A moves to the left. The clearance between A and B is greater at the left than at the right. There will be a tendency for more fluid to enter at the left than leaves at the right. Thus, a pressure is developed in the oil which serves to separate the surfaces.

The situation for a journal bearing is shown in Fig. 12-5. As the journal A rotates counterclockwise it tends to drag the oil that adheres to its surface into the narrowing wedge, and the same action as described for Fig. 12-4 takes place. The resultant of the pressure produced raises the shaft and displaces it slightly to the right.

The formation of a film will take place only when the proper relation exists between a number of factors. Some of these factors are the speed at which the journal rotates, the viscosity of the oil, and the magnitude of the load relative to the size of the bearing. The coefficient of friction varies from about 0.002 to 0.02, increasing with an increase in ZN/p.

If the load is too great, or if the speed or viscosity is too small, a thick-film condition will not exist. Even when the normal operating conditions are such that a thick film forms, there will be short periods, during starting and stopping, when the thick film will not exist, and the situation will be one of imperfect lubrication.

12-8 BOUNDARY LUBRICATION

It was stated earlier that a very thin film quickly forms on a bare metal surface and that, because of this film, friction and wear are reduced. Though this is boundary lubrication, the term is generally used for situations where attention is given to causing the formation of a particular kind of film. With such a film on one or both surfaces the friction is greatly reduced. Under favorable conditions the coefficient of friction may be as low as 0.04, though a value twice this would be a more reasonable minimum, and the maximum would approach the value with no intentional lubrication, which for metal on metal averages 0.30.

A number of substances are used to cause the formation of these films, including the fatty acids in animal or vegetable oils, solids such as graphite or molybdenum disulfide, and grease. If mineral oil is used, a small amount of one of the fatty acids should be added.

The material of the surfaces is also a factor, for the formation of some of the films requires a chemical reaction with the metal to take place in order to produce the film. Also, there will be instances where the film is removed and the surfaces will weld together. These welds must break as motion continues, or relatively large particles of the surface will be broken out. Therefore, metals must be selected which produce welds that are easily broken.

It must be remembered that, though these substances which produce a thin film are effective in reducing friction and wear, they will not remove heat.

12-9 IMPERFECT LUBRICATION

When a bearing is operating in the realm of imperfect lubrication, the surfaces are not separated by a continuous thick film. However, because of the configuration of the surface — due to surface roughness — there are many small pockets of oil that tend to develop a supporting pressure in the manner described for perfect lubrication. Thus, part of the load forcing the surfaces together is supported by a film of oil, and part is supported in the manner of boundary lubrication.

The proportion of the load supported in each of these ways varies with the conditions under which the bearing is operating. The coefficient of friction depends on the manner in which the load is divided, approaching the perfect lubrication value at one extreme and the boundary lubrication value at the opposite extreme. Thus, the range is about 0.02 to 0.10.

There is a theory upon which to base the design of bearings with perfect lubrication, but not for imperfect or boundary lubrication. Their design must be based on empirical data.

12-10 THRUST BEARINGS

The simplest form of thrust bearing is in the form of a large flat washer that is made of, or faced with, a bearing material. It is placed in contact with the end of the shaft or with a collar on the shaft. Several radial grooves are often cut in the surface, as shown in Fig. 12-6a, to distribute the lubricant. This bearing does not develop a thick film and should be limited to slow speed or intermittent operation and to bearing pressures of less than 100 psi.

A tapered-land thrust bearing has the general appearance of the bearing just mentioned, except that the surface is so machined that a thick film is developed in the manner described in Sec. 12-7. A cross section of one of the segments of the surface is shown in Fig. 12-6b. The permissible bearing pressure is several times that of the above bearing. Another form of thrust bearing that operates in the realm of perfect lubrication is the

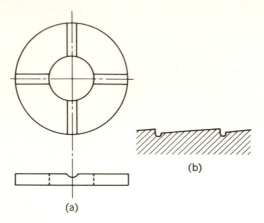

Fig. 12-6   A thrust bearing.

tilting-pad type. Instead of the segments being machined to provide the required taper, they are made as separate parts and so designed that they pivot to provide the taper, as shown in Fig. 12-7. Thrust bearings of this type have been made that operate with a load of 3 million lb.

12-11   SLIDING SURFACES

Examples of sliding surfaces are those machines in which a portion, such as a table, moves and is supported and guided by the base, as is the case with planers and shapers. A thick film cannot be developed in this type of bearing, and lubrication is poor. Thus, the allowable bearing pressures must be reduced. For cast iron on cast iron, the maximum pressure should be about 75 psi.

The surfaces involved are usually on a large part and, therefore, cannot be replaced when they become worn. So wear must be provided for in

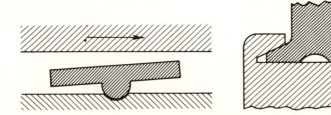

Fig. 12-7   A tilting-pad thrust bearing.   Fig. 12-8   Provision for wear.

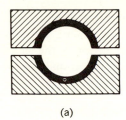

(a)

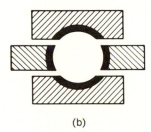

(b)

Fig. 12-9 Journal bearings designed to permit compensating for wear.

some other manner. Figure 12-8 is one method. The situation is often complicated by a greater amount of motion — therefore, greater wear — taking place in a small part of the total travel. In such a case, the only provision that can be made is to make it as easy as possible to remachine the worn parts.

12-12 JOURNAL BEARINGS

The simplest form of journal bearing is a hole drilled in a piece of material that has desirable bearing characteristics and properties. Many other configurations are found for one or more of the following reasons:

1. To achieve perfect lubrication
2. To compensate for wear by adjusting segments of the bearing or by removing shims (Fig. 12-9)
3. To overcome difficulties associated with certain applications, such as vibration, which can be produced by the bearing
4. To provide for proper imperfect lubrication
5. To facilitate economical manufacture of the bearing

A full bearing intended to operate satisfactorily with imperfect lubrication is by far the most often used plain bearing and, therefore, will be discussed in detail in this chapter.

12-13 BEARING MATERIALS

In selecting a bearing material, the type of load and speed are obvious factors, but also important are factors such as the environment, the type and reliability of lubrication, and the alignment and deflection of the shaft. The material properties and characteristics to be considered in making the selection are as follows:

1. Strength — The bearing pressure that may be permitted depends on the bearing strength or the compression strength. It must be kept in mind that most metals suitable for use as bearings begin to lose their strength at relatively low temperatures.
2. Hardness — Greater strength is accompanied by greater hardness, and a harder bearing material requires a harder shaft.
3. Compatibility — When the two surfaces of a bearing are in bare metal-to-metal contact, they weld together, and there is a tendency to destroy the surfaces. The magnitude of this tendency is a measure of the compatibility of the two metals. For example, a hardened steel shaft and a soft aluminum bearing are compatible, but a shaft and bearing both of aluminum are not.
4. Conformability — A material that possesses good conformability compensates for slight misalignment, deflection, and tolerances. A low modulus of elasticity is an indication of good conformability.
5. Embeddability — A material possessing good embeddability allows grit and other foreign particles to become embedded, thus preventing the damage to the journal that would otherwise occur. Usually, metals with good conformability also have good embeddability.
6. Cost — The cost of bearing materials varies greatly. It must be remembered that the cost of the bearing material is small compared with the cost of bearing failure, even if just the cost of replacing it and the time that is lost is considered. There may also be considerable damage to the shaft.
7. Fatigue resistance — If the load is steady, fatigue resistance is of little importance. But if there are many stress cycles, it may be the most important characteristic of the material. Fatigue resistance as applied to bearing materials is different from the usual fatigue resistance. It decreases with increase in temperature.
8. Corrosion resistance — This term applies to the resistance to the corrosive agents that may develop in the lubricant. The corrosion rate increases with an increase in temperature.
9. Effects of temperature — The effect of temperature on strength, fatigue, and corrosion has been mentioned. It is the temperature at the surface of the bearing that is important. Thermal conductivity is especially important where there is not a flow of lubricant through the bearing.

All the desirable properties and characteristics cannot be had in any one material. Thus, any choice involves a compromise based on the relative importance of the various factors and their effect on the material. A great many bearing materials have been developed. Here, it will be possible to discuss only those must often used. Their characteristics are given in Table 12-1, using an arbitrary scale from A to E, where A is the most desirable.

Tin and lead babbitts are very common bearing alloys. Tin babbitts have about 80 percent tin, and lead babbitts have about 80 percent lead. The load-carrying capacity of babbitts is lower than that of most other bearing materials, and their strength decreases greatly with an increase in temperature. They are widely used, because they satisfy most requirements for general applications. Lead babbitts are cheaper than tin babbitts, and are the most often used. SAE 15 is the most popular lead babbitt.

Copper alloys are often used in bearings. They have higher load-carrying capacity, greater wear resistance, and are better suited to high temperature operation than babbitts. However, Table 12-1 shows that they are inferior in other respects. Leaded bronzes are generally more satisfactory than tin bronzes. The most often specified bronze for cast bearings is SAE 660.

Aluminum alloys developed for bearing applications are finding increased use, because of their high load-carrying capacity and low cost. The alloy that is most common is SAE 770.

Silver has some very desirable characteristics, but the precision and rigidity required limit its use.

The properties of these metals when used as bearings are given in Table 12-2.

12-14 DETERMINING BEARING SIZE

A number of interrelated factors are involved in the determination of the size of a bearing that is required for a particular application. There

TABLE 12-1 Characteristics of Bearing Materials

MATERIAL	COMPAT- IBILITY	CONFORM- ABILITY	EMBED- DABILITY	FATIGUE RESIS- TANCE	CORROSION RESIS- TANCE	THERMAL CONDUC- TIVITY
Tin-base babbitt	A	B	A	D	A	D
Lead-base babbitt	A	B	B	D	C	D
Lead bronze	C	D	D	B	C	C
Aluminum	C	C	D	B	A	C
Silver	C	E	D	A	A	A

TABLE 12-2 Properties of Bearing Materials

MATERIAL	MAXIMUM PRESSURE, PSI	MAXIMUM VELOCITY, FPM	MAXIMUM TEMPERATURE, °F
Tin-base babbitt	1500	4000	250
Lead-base babbitt	1200	4000	300
Lead bronze	3500	1000	400
Aluminum	4000	1500	250
Silver	4000	4000	300

is also a very wide range of operating conditions. The cycle of operation greatly affects bearing size. The bearing may be required to operate continuously under full load for long periods, or the full load may be applied only for very brief periods. Bearing pressure is the load divided by the projected area, and is represented by p. The extreme range is from 1 psi for cotton mill spindles to 10,000 psi for slowly moving turntables. An idea of the normal range of pressures may be gained from Table 12-3.

Speed also affects the size of a bearing and is dictated by consideration of factors other than those pertaining to bearings. The speed in revolutions

TABLE 12-3 Bearing Design Data

APPLICATION	BEARING	MAX. PRESSURE, PSI	MIN. ZN/P	Z AT 100°F	L/D RANGE
Electric motors, generators, centrifugal pumps	Main	200	200	30	1.0–2.0
Reciprocating pumps and compressors	Main	250	30	80	1.0–2.0
	Crankpin	600	20	80	1.0–1.7
	Wristpin	1000	10	80	1.5–2.0
Gasoline engines	Main	1800	15	125	0.6–1.6
	Crankpin	2700	10	125	0.6–1.2
	Wristpin	4000	8	125	1.4–2.0
Punch press	Main	4000	2	100	1.0–2.0
	Crankpin	7000	2	100	1.0–2.0

per minute is not as important as the relative velocity of the sliding surfaces. This velocity is found with the equation

$$V = \frac{\pi DN}{12} = 0.262DN \tag{12-2}$$

where V = velocity, ft per min (fpm)
D = journal diameter, in.
N = rpm

It should be noted that the velocity increases with the journal diameter, as well as with the revolutions per minute.

Velocity and permissible pressure are very closely related. The factor pV is often used in determining the size of bearings, especially those with imperfect lubrication. The factor ZN/p is also used in determining the size of bearings; values for various applications are given in Table 12-3.

It would appear that sufficient area could be provided by merely making the bearing long enough, and to an extent this is done. However, there are limitations. The permissible misalignment is reduced by longer bearings. Longer bearings are adversely affected to a greater degree by shaft deflection. Longer bearings also require greater space and cause an increase in weight. The recommended ratio L/D, where L is the bearing length and D the diameter, is given in Table 12-3.

It should be noted that the pressures in Table 12-2 are the maximum for the material. Those in Table 12-3 are operating pressures for the particular application. The viscosity in Table 12-3 is the recommended viscosity at 100°F. The operating temperature probably will be higher than this, and the operating viscosity will be lower. The viscosity at 100°F is given to aid in selecting a suitable oil.

Example 12-2   What is the length of a bearing for a 2-in.-diameter journal, if the velocity is 210 fpm, the viscosity is 30 centipoises, the load is 2400 lb, and the ZN/p is 20?

Solution   Rearranging Eq. (12-2),

$$N = \frac{V}{0.262D} = \frac{210}{0.262 \times 2} = 400 \text{ rpm}$$

ZN/p is 20; therefore

$$20 = \frac{30 \times 400}{p}$$

$$p = 600 \text{ psi}$$

$$\text{Area required} = \frac{2400}{600} = 4 \text{ sq in.}$$

Therefore, the 2-in.-diameter bearing must be 2 in. long.

Fig. 12-10   Split strip-type bearing.
(*Federal-Mogul Corp.*)

12-15   BEARING CONSTRUCTION

There are several ways in which a simple journal bearing can be made. Strip-type bearings are made from sheet metal. This sheet is generally steel to which a bearing metal has been bonded. They may be either full cylindrical or split, as shown in Fig. 12-10. This kind of bearing must be made in large quantities, because of the tooling involved. They are available in several types and in many sizes.

A lining of bearing metal can be applied to a part that is made from a casting or forging. This method of providing a bearing is becoming less common. A bearing may also be machined from a casting or tubing made of a bearing metal. Bearings of this kind are available in many standard types and sizes.

12-16   DESIGN DETAILS

The successful performance of a bearing is dependent, to a considerable extent, on the attention given to the details of the design associated with it. When it is desired to take advantage of hydrodynamic action, oil inlet holes and distribution grooves must be located where the oil pressure is at a minimum. The axial location for oil holes in horizontal bearings should be at the center, and in vertical bearings it should be near the upper end. The end of the shaft should not be inside the bearing, a groove should not be cut in the shaft to distribute oil, and the bearing should not be placed in a blind hole. These practices interfere with the required oil flow.

In the case of imperfect lubrication, the location of the lubricant distribution grooves is not critical, but they should not be excessive, because of the reduction of area caused by grooves. Also, they should not break through the ends of the bearing. As much area as possible should be provided between the bearing and the housing to facilitate heat transfer.

The desired clearance (difference between journal diameter and bearing diameter) where hydrodynamic action is intended is 0.001 in./in. of diameter. Where imperfect lubrication prevails this clearance is 0.002 in./in. of diameter.

The surface roughness of the bushing should not be greater than 32 μin. and the journal should be 20 μin. or smoother. The higher the speed, or the greater the pressure, the more important a smooth surface becomes. The hardness of the shaft should be not less than Brinell 120 for babbitt bearings, 300 for bronze, and 250 for aluminum. If the bearing is to be a press fit in the hole, and the intention is not to machine it after assembly, an allowance for the reduction of the inside diameter must be made. For machined bearings, the reduction will usually not be less than three-fourths of the interference and may equal the total interference, depending on the configuration and material of both the part and the bearing. A chamfer should be provided on the part to facilitate inserting the bearing, and it should be a 45° included angle, rather than a 45° chamfer. The size of the chamfer along the axis of the hole should be about 10 times the maximum interference.

12-17 POROUS BEARINGS

Porous bearings, also called sintered metal bearings or powder-metal bearings, are generally bronze, though iron or combinations such as iron and copper are sometimes used. Only bronze will be treated here. An advantage of this kind of bearing is that the lubricant is contained within the bearing. This feature often permits a simpler design.

The permissible pV factor under favorable conditions is 50,000, though a value of about one-half this would be more appropriate where life and reliability are important, and under severe conditions it must be further reduced. For thrust bearings, the permissible pV is 10,000. Velocities above 1500 fpm require special consideration and a reduced pV.

Though the oil contained in the bearing is generally sufficient for the life of the bearing, supplementary lubrication may be desirable. In such cases, provision should be made to supply oil to the outside diameter or end of the bearing. Oil is preferred. However, grease may be used, if it is applied directly to the journal rather than to the outside diameter or end of the bearing.

12-18 MISCELLANEOUS BEARING MATERIALS

Cast iron can be used for a bearing if attention is given to alignment, surface roughness, and lubrication. The pressure should be less than 650 psi and the maximum velocity 130 fpm. Rubber is an excellent choice where water makes oil lubrication difficult, such as for marine propeller shaft bearings. The presence of sand in the water used as the lubricant has little effect, because of the embeddability of the rubber. The pressure should be limited to 50 psi, and the temperature to 150°F. A pV less than 15,000 is recommended. The bearing must not be permitted to run dry. Nylon may be used to advantage in some applications, even without lubrication. The pressure limit is 1000 psi and the pV is 2500. Carbon graphite is used in special applications where oil or grease lubrication is undesirable, such as for food-processing machinery. The maximum pressure should be less than 50 psi, and the pV dry should be about 15,000. The temperature may be 900°F when oxygen is present, and several times this if oxygen is not present.

QUESTIONS

12-1   What are the two basic types of plain bearing?

12-2   What is kinetic friction?

12-3   What is meant by "wearing in" a bearing?

12-4   What are three major reasons for using a lubricant in bearings?

12-5   What is the most important property of oil with respect to lubrication?

12-6   What is a Saybolt universal second?

12-7   What is meant by the term boundary lubrication?

12-8   What is the difference between hydrostatic lubrication and hydrodynamic lubrication?

12-9   What is imperfect lubrication?

12-10 What is a full-journal bearing?

12-11 What is the difference between compatibility and conformability as applied to bearing materials?

12-12 What is embeddability?

12-13 What is the most desirable characteristic of babbitt as a bearing material?

12-14 What is the most desirable characteristic of aluminum as a bearing material?

12-15 What is the least desirable characteristic of silver as a bearing material?

12-16 What are some of the disadvantages of a long bearing?

12-17 What is a strip bearing?

12-18 What is the main advantage of a porous bearing?

12-19 Under what conditions is rubber used as a bearing material?

12-20 What are the advantages of carbon graphite as a bearing material?

PROBLEMS

Note: Upon completion of some of these problems, Design Example 3 and Project 8 or 9 in Chap. 19 are appropriate.

12-1   An oil has a viscosity of 2 stokes. What is its viscosity in centipoises?*
12-2   An oil has a viscosity of 300 SUS. What is its viscosity in centipoises?
12-3   If the viscosity is 400 SUS, the diameter of the bearing is 3 in., the length is 3.5, the load is 3000 lb, and the rpm is 250, what is the ZN/p?*
12-4   A 3.5-diameter bearing that turns 275 rpm is to operate at 150°F. SAE 20 oil will be used, and the ZN/p is to be 90. What is the length of the bearing?
12-5   What is the velocity in feet per minute for a 3.75-diameter journal, running at 200 rpm?*
12-6   If a 4.25-diameter bearing has the velocity limited to 3800 fpm, what is the rpm?
12-7   A rubber bearing 4 in. in diameter and 6 in. long is to carry a load of 1000 lb at 500 rpm. Is the size adequate?
12-8   A 2.375-diameter bearing with an L/D of 1.7 turns 1800 rpm, and supports a load of 200 lb. What is the pV factor?*
12-9   A nylon bearing 0.4 in. in diameter has an L/D of 2 and carries a load of 16 lb. What is the maximum rpm?
12-10 The pV is to be 6000, the diameter is 2.75, the length is 3.25, and the load is 380 lb. What is the maximum rpm?

SELECTED REFERENCE

Wilcock, D. F., and E. R. Booser: "Bearing Design and Application," McGraw-Hill Book Company, New York, 1957.

13 BALL AND ROLLER BEARINGS

13-1 INTRODUCTION

Bearings in which rolling is involved rather than sliding are to be discussed in this chapter. A comparison of various kinds of bearings will also be made.

There are two basic types of rolling bearings: those that use a ball as the rolling element and those that use a roller, which may be either cylindrical or tapered. Rolling bearings could also be divided into two classes on the basis of whether they are radial or thrust bearings.

13-2 BALL BEARING TERMINOLOGY

The construction of a ball bearing is illustrated in Fig. 13-1. The parts and dimensions are identified in Fig. 13-2: (*a*) the outer race; (*b*) the inner race; (*c*) one of the balls; and (*d*) the retainer or cage. The dimensions are: (*e*) the diameter; (*f*) the bore; and (*g*) the width.

13-3 ROLLING BEARING THEORY

Though the following discussion will refer to ball bearings, the concepts also apply to roller bearings.

Theoretically, a ball makes point contact with the race, and when a load is applied, deformation of the ball and the race takes place to the extent necessary to provide sufficient area to support the load.

If a ball is subjected to a load applied by means of two balls, point contact is approached, and the ball will support a certain load at a given stress in the ball.

Fig. 13-1 Ball bearing construction. (*SKF Industries.*)

If the load is applied by grooves whose radius is twice the ball diameter, the ball can support 16 times as much load as when loaded by means of the balls before exceeding the given stress. The grooves in a ball bearing have a radius slightly greater than the radius of the ball.

The load-carrying capacity of a ball varies as the square of its diameter. Thus, the larger the balls in a bearing, the greater its load capacity. However, if the thickness of the races is made too small in an effort to maintain a certain bearing diameter, the life of the bearing will be adversely affected.

The small amount of friction is one of the main advantages of a ball bearing. The reduction in static friction is particularly pronounced, and thus the surging that takes place in a sliding bearing when the static friction is overcome is practically eliminated. The actual friction for all practical purposes is negligible. Often the friction of the seals is greater than that of the bearing itself.

A major source of friction is the rubbing between the balls and the

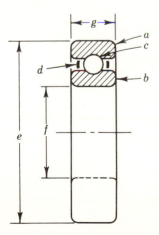

Fig. 13-2 Ball bearing terminology.

separator. The function of the separator is not only to space the balls so as to properly distribute the load, but to prevent the balls from rubbing on each other. If the balls were not separated, their surfaces with the greatest velocity would come in contact, and these surfaces are moving in opposite directions. The extremely smooth surface finish characteristic of ball bearings is not produced to reduce friction, but it is the result of the accuracy required in their manufacture.

A ball bearing of an appropriate size for the application, that is properly installed and maintained, will fail as a result of surface fatigue; that is, a very small flake of metal will be removed from one of the races or balls, probably the inner race. The bearing has not ceased to operate. Nevertheless, it must be regarded as failed and must be replaced at this time if a continuation of good performance is desired. Fatigue is a function of the number of applications of the stress. Thus, the life of a ball bearing is so many revolutions at a specified load.

13-4 RADIAL BALL BEARINGS

Radial ball bearings are the most used of any of the rolling bearings, and the plain single-row type, also called Conrad type, is the most popular. Such a bearing is shown in Fig. 13-1, and a typical cross section is shown in Fig. 13-2. These bearings can carry a radial load and a small thrust load. A Conrad bearing is assembled by placing the inner and outer races together so that they form a crescent-shaped space on the opposite side, into which the balls are placed. When the greatest number of balls has been inserted, they are spaced, and the two parts of the retainer are joined together, usually by riveting.

The Conrad type is available with the inner race extended beyond the width of the outer race on one or both sides. This type is also available with

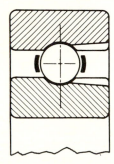

Fig. 13-3 A filling-slot bearing.

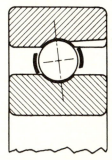

Fig. 13-4 An angular-contact bearing.

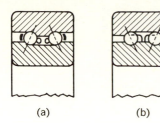

(a)   (b)   **Fig. 13-5   Double-row   angular-contact   bearings.**

the outer race grooved to receive a snap ring, which simplifies assembly in some instances. The snap ring will withstand a load greater than the thrust rating of the bearing.

Because of the manner in which the Conrad type is assembled, a limited number of balls may be inserted. If a larger number of balls could be used, the radial load could be increased. In the filling-slot type, a semicircular slot is cut in both the inner and outer race to permit adding additional balls, as shown in Fig. 13-3. These slots do not extend to the bottom of the grooves, and thus do not interfere with the contact area developed when carrying a radial load. However, the filling slots do interfere with the contact area developed when subjected to a moderate thrust load. Thus, if the thrust load is more than half the radial load, the Conrad type is a better choice. This type is not suitable for a pure thrust load.

An angular-contact bearing is one that has deep grooves and has one shoulder of the outer race almost completely removed to permit assembly. An example of this type is shown in Fig. 13-4. This design permits inserting a large number of balls. Thus, it can carry a large thrust load or combination of thrust and radial loads. However, it can take thrust in only one direction, and when mounted singly cannot take a pure radial load. Bearings of this type that are slightly modified and intended to be used in pairs, called duplex pairs, will be discussed in Sec. 13-10.

An angular-contact bearing known as type J has one shoulder on the inner race partially removed. This type is particularly well suited to high-speed applications. Another type of angular-contact bearing, sometimes called a "magneto" bearing, has one shoulder of the outer race completely removed, which permits the bearing to be readily assembled or disassembled.

Double-row angular-contact bearings are, essentially, two single-row bearings in a single unit. This type is capable of carrying a large radial load or a large thrust load from either direction. There are two types: the externally converging type, shown in Fig. 13-5a, where the contact angles converge outside the bearing, and the internally converging type, shown in Fig. 13-5b, in which the angles converge inside the bearing. The externally converging type resists misaligning forces, while the internally converging type can accept moderate misalignment, such as shaft deflection.

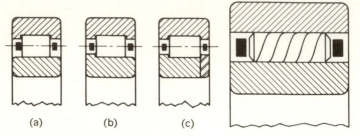

Fig. 13-6   Cylindrical roller bearings.   **Fig. 13-7   A wound roller bearing.**

13-5   RADIAL ROLLER BEARINGS

Roller bearings are divided into two basic kinds: those that have cylindrical rollers and those that have tapered rollers. There are many types of cylindrical roller bearings. The plain single-row type may be nonlocating, as shown in Fig. 13-6a. This provides freedom of axial movement of the shaft in either direction. The one-direction locating type shown in Fig. 13-6b allows axial movement in one direction. Both are readily separated, which facilitates assembly and disassembly of the machine of which they are components. Also available is the type shown in Fig. 13-6c, which has two ribs. Cylindrical roller bearings are primarily intended to carry heavy radial loads. However, those with ribs can carry a thrust load, but it should not exceed 0.2 of the radial load. It should be noted that this thrust load is carried by sliding, and thus there will be considerable friction.

Another type, known as a journal roller bearing, has rollers that are long relative to their diameter. They do not have ribs and, therefore, cannot carry a thrust load. They are available without one or both races; the performance then is dependent on the hardness and roughness of the surfaces that take the place of the race.

A unique form of cylindrical roller bearing is shown in Fig. 13-7. The rollers are made by winding narrow steel strips in the form of a helix. At assembly of the bearing the direction of winding is alternated. This form of roller construction tends to sweep the lubricant back and forth across the surfaces. It also allows for slight irregularities in the surface, which is desirable when used without a race. It is well suited to heavy shock loads but not to high speeds.

Tapered roller bearings can carry both large radial loads and large thrust loads. For a single-row bearing, the thrust load can be only in one direction, and a pure radial load cannot be carried. These bearings employ a terminology different than that for other bearings. The terminology is shown in Fig. 13-8: (a) the cup; (b) the cone; (c) one of the tapered rollers; (d) the cage; (e) the back face of the cone; (f) the cup front face; and (g) the bearing width.

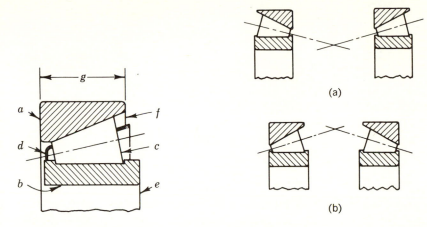

Fig. 13-8 Tapered roller bearing termi- **Fig. 13-9 Tapered roller bearing ar-**
nology. **rangements.**

When a radial load is applied to a single-row tapered roller bearing, a radial component and a thrust component are produced. The tapered surfaces take all the radial load and most of the thrust load. The remainder of the thrust load is carried by the rib at the back face of the cone. This sliding of the roller against the rib increases the friction. The coefficient of friction is roughly twice that of a cylindrical roller or ball bearing.

Tapered roller bearings are readily separated. Added attention must be given to alignment, and they are not recommended for high-speed applications. They are available with the rollers inclined at various angles; the steeper the angle, the greater the thrust load that can be carried, but the radial load is reduced.

A pair of single-row tapered roller bearings may be arranged in either of the ways shown in Fig. 13-9. The arrangement in Fig. 13-9a provides a greater rigidity and stability than that in Fig. 13-9b. It also requires closer alignment than does the arrangement in Fig. 13-9b, and the installation design is usually more complicated and thus more costly.

Either of the arrangements shown in Fig. 13-9 is available as a double-row bearing. Four-row bearings and a number of other special configurations are also available.

13-6 BEARING CAPACITY

A bearing's capacity involves its capability in respect to load, speed, and life, all of which are interrelated. There is an exponential relationship between the magnitude of the load on a bearing and its life. It is generally accepted that if the load is doubled, the life is reduced to one-eighth, or if the load is halved, the life is increased 8 times. The relationship between load, life, and capacity is

$$L_n = \left(\frac{C}{P}\right)^3 \qquad\qquad (13\text{-}1)$$

where L_n = life, millions of revolutions
$\qquad C$ = specific dynamic capacity, lb
$\qquad P$ = radial or equivalent load, lb

The specific dynamic capacity C is the load that 90 percent of a group of bearings can carry for 1 million revolutions before the first evidence of fatigue develops. Some manufacturers use what they call the average life, for which 50 percent rather than 90 percent is used.

Table 13-1 contains a representative specific dynamic capacity for a few selected bearings, for use primarily in examples and problems. The

TABLE 13-1   Specific Dynamic Capacity of Selected Bearings (in pounds)

BORE NO.	10 SERIES	03 SERIES	73 SERIES	N3 SERIES
01	800	1,350	1,550	
05	1,650	3,650	4,750	5,200
06	2,200	4,800	6,300	6,800
07	2,650	5,700	7,950	8,800
08	2,850	6,950	9,900	10,400
09	3,550	9,000	12,000	14,300
10	3,750	10,400	16,300	17,000
11	4,800	12,000	18,900	20,800
12	5,100	13,700	21,700	23,200
13	5,300	16,000	24,200	26,000
14	6,550	18,000	24,700	30,500
15	6,950	19,600	30,800	33,000
16	7,350	21,300	32,000	36,500
18	10,000	23,000	37,500	47,500
20	10,500	30,000	43,000	63,000

10 and 03 series are Conrad type, the 73 series is an angular-contact type, and the N3 series is a cylindrical roller bearing. The sizes of these bearings are given in Table 13-2.

Example 13-1   Select a 10 series bearing to carry a 3500-lb radial load and have a life of 1.5 million revolutions.

TABLE 13-2 Dimensions of Selected Bearings (in inches)

BORE NO.	BORE	10 SERIES		03, 73, N3 SERIES	
		DIAMETER	WIDTH	DIAMETER	WIDTH
01	0.4724	1.1024	0.3150	1.4567	0.4724
05	0.9842	1.8504	0.4724	2.4409	0.6693
06	1.1811	2.1654	0.5118	2.8346	0.7480
07	1.3780	2.4409	0.5512	3.1496	0.8268
08	1.5748	2.6772	0.5906	3.5433	0.9055
09	1.7716	2.9528	0.6299	3.9370	0.9842
10	1.9685	3.1496	0.6299	4.3307	1.0630
11	2.1654	3.5433	0.7087	4.7244	1.1417
12	2.3622	3.7402	0.7087	5.1181	1.2205
13	2.5590	3.9370	0.7087	5.5118	1.2992
14	2.7559	4.3307	0.7874	5.9055	1.3780
15	2.9528	4.5276	0.7874	6.2992	1.4567
16	3.1496	4.9213	0.8661	6.6929	1.5354
18	3.5433	5.5118	0.9449	7.4803	1.6929
20	3.9370	5.9055	0.9449	8.4646	1.8504

Solution   Rearranging Eq. (13-1),

$$C = P \sqrt[3]{L_n}$$

$$C = 3500 \sqrt[3]{1.5} = 4000 \text{ lb}$$

From Table 13-1, a No. 11 bore bearing with a specific dynamic capacity of 4800 lb is selected.

There is a maximum permissible speed at which a bearing should be operated. This limit involves the factor DN, in which D is the bore of the bearing in millimeters and N is the revolutions per minute. Including a factor which allows for the diameter in inches, the expression becomes $25.4DN$. The maximum permissible DN is dependent not only on the type of bearing, but on its internal design and the material of the retainer. Representative values are: radial ball bearings — 500,000, double-row bearings — less, and angular-contact bearings with a small angle — considerably higher; radial roller bearings — 300,000, and cylindrical roller thrust bearings — only 100,000.

In the usual situation, the load is applied when there is motion between the races of a bearing. However, the load may be applied when there is no motion. This is referred to as a static load. When an extremely high static load is applied, very small dents are left in the race, which cause vibration

TABLE 13-3 Static Capacity of Selected Bearings (in pounds)

BORE NO.	10 SERIES	03 SERIES	73 SERIES	N3 SERIES
01	450	800		
05	1,100	2,500	2,850	3,900
06	1,500	2,800	3,800	5,300
07	1,900	3,550	4,550	7,100
08	2,100	4,500	5,700	8,300
09	2,700	5,500	7,650	11,400
10	2,900	6,650	9,150	14,300
11	3,800	7,900	10,800	17,000
12	4,050	9,250	12,500	19,000
13	4,400	10,650	14,300	21,200
14	5,400	12,200	16,300	25,000
15	5,850	13,100	18,300	30,500
16	6,950	14,800	20,400	33,000
18	8,650	18,500	25,500	39,000
20	9,300	24,650	34,500	53,000

and noise when the bearing is subsequently rotated. Table 13-3 gives a representative static capacity for the selected bearings.

A bearing can carry a load several times this limit when slowly rotating. The load may be 4 times this limit, if the increase in noise and decrease in life are not objectionable. Thrust loads greater than the limiting static thrust rating should not be imposed on rotating bearings.

13-7 BEARING SIZE DETERMINATION

A number of factors must be taken into account when determining the size of a bearing for a particular application. The design load is calculated by applying appropriate values for strength factors b and d to the working load, as discussed in Sec. 7-7.

A bearing is often subjected to both radial and thrust loads, and this combination must be converted to an equivalent load. The equivalent load depends on the manufacturer as well as on the type of bearing. However, a rough, though conservative, approximation can be made with

$$P_e = 0.5R + 1.7T \tag{13-2}$$

where P_e = equivalent radial load, lb
$\quad\quad R$ = radial load, lb
$\quad\quad T$ = thrust load, lb

If this equation produces an equivalent load less than the radial load, the radial load should be used as the equivalent load.

Example 13-2 Select an 03 series bearing to carry a radial load of 4800 lb and a thrust load of 1100 lb, and have a life of 75 hr at 800 rpm.

Solution The total number of revolutions is equal to

$$75 \times 60 \times 800 = 3.6 \text{ million}$$

The equivalent radial load is found with Eq. (13-2):

$$P_e = (0.5 \times 4800) + (1.7 \times 1100)$$

$$P_e = 2400 + 1870 = 4270 \text{ lb}$$

Since this is less than the radial load, 4800 is used as the equivalent load. Rearranging Eq. (13-1),

$$C = P \sqrt[3]{L_n}$$

$$C = 4800 \sqrt[3]{3.6} = 7350 \text{ lb}$$

From Table 13-1 a No. 09 bore bearing is selected.

A smaller contact area is developed in the inner race of a bearing because of its sharper curvature, also because this curvature is opposite to that of the rolling element. Because of this smaller contact area, the stress in the inner race is greater than in the outer race. When the inner race rotates, as is the usual case, the same spot in the outer race is continuously loaded, but as the stress is less than in the inner race this is not critical. However, when the inner race does not rotate, the load is continuously applied to the same spot on the inner race. To allow for this, the radial load is multiplied by 1.2.

It should be noted that there is as yet no universally accepted method of specifying the capacity of a bearing and calculating the permissible load or probable life for a specific application. Therefore, the method of the manufacturer must be used with his ratings.

13-8 BEARING SHIELDS AND SEALS

Some bearings are available with shields or seals as part of the bearing assembly. Their purpose is to protect the bearing. Shields are generally permanently attached to the outer race and fit a notch in the inner race with a slight clearance. They exclude all but the finest grit. Bearings are available with a shield on one or both sides. Figure 13-10 shows a cross section with a shield on one side.

A seal provides protection from all contaminants, because it rubs against the race. It may also retain the lubricant placed in the bearing by the manufacturer. A seal may be used on one or both sides. Bearings are

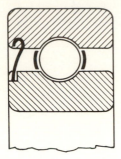

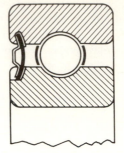

Fig. 13-10 A ball bearing with shield.   **Fig. 13-11 A ball bearing with seal.**

also available with a shield on one side and a seal on the other. Figure 13-11 shows a seal on one side.

Because a seal rubs on the race, there is added friction. The amount of power lost because of this friction is usually negligible. However, the heat generated at high speed is not.

Shields and seals are most often used on ball bearings. Some cylindrical roller bearings are available with them, though most applications dictate that the bearing be open to permit the lubricant to remove the heat generated. When a seal is desirable but is not available as part of the bearing, or when it is undesirable as part of the bearing, a separate unit may be used. Seals of this kind are discussed in Chap. 18.

13-9 LUBRICATION

The principles involved in the lubrication of rolling bearings are greatly different than those which apply to sliding bearings. Because of the extremely high pressure between the rolling element and the race, any lubricant is squeezed out and there is metal-to-metal contact. A lubricant is, nevertheless, essential to the operation of a rolling bearing for the following reasons:

1. To reduce the friction at the points where sliding takes place
2. To control the heat caused by friction and deformation
3. To prevent corrosion
4. To aid in protecting the bearing from dirt

Where removal of heat is not involved, a fraction of a drop of oil per hour will provide adequate lubrication for a moderate-size bearing, if properly applied. Considering only lubrication, oil is best, but grease is used to a greater extent, because it permits a simpler installation design and the problem of leakage is practically eliminated. Grease should not be used where the DN is in excess of 200,000, unless special precautions are taken.

In applications where cooling is not a major function of the oil, the distribution caused by a part splashing through a pool of oil is usually adequate. If the lower portion of a bearing is to rest in a pool of oil, the surface of the oil when the machine is not operating should be about at the center of the lowest ball or roller.

It would be an error to conclude that, because of the great latitude allowed in lubricating rolling bearings and the small amount that is necessary, lubrication is not important. Lack of lubrication will usually cause destruction of the bearing, and this can result in major damage to the machine of which it is a part.

13-10 BEARING INSTALLATION DESIGN

Unless the installation design adequately provides for the bearing, its precision is largely wasted. Generally, the rotating race is made a firm interference fit, and the stationary race fit is made loose enough to permit a very slight rotational creep. This rotational creep will avoid prolonged stressing of the same spot on the race. When a press fit is used, there will be a slight decrease in the internal clearance of the bearing. The fits specified by the manufacturer should be adhered to.

Whenever there is a difference in the thermal expansion of the shaft and housing, the difference must be allowed for in the selection of a bearing type or in the mounting. Some machine elements have a tendency to align themselves and maintain this alignment, and in some instances it is desirable to permit this. There are three methods of mounting a shaft. First, the shaft is not located axially by the bearings. Second, the shaft is located axially by a bearing at one end that takes thrust in both directions; the bearing at the other end is of a type that allows axial movement, such as a nonlocating roller bearing. Third, the shaft is carried by bearings at each end which can take thrust in only one direction but which are mounted opposed to provide for thrust in either direction. The second method should be used to provide for differences in thermal expansion, and the bearing that takes the thrust should be located close to the point where axial location is most important.

Attention should be given to the surfaces that serve as bearing seats. The diameters should be perfectly round, the shoulders square with the diameters. The surfaces should have a smooth finish, preferably ground. The same degree of accuracy applies to parts that serve to clamp bearings, such as gears or spacers. Shoulders should be low enough to permit removal of the bearing and large enough to take any thrust load that is in-

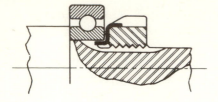

Fig. 13-12 A means of clamping the inner race.

volved. The fillet radii must be less than the matching corner radius on the bearing. A relief at the shoulder will permit a large fillet without interfering with the bearing, and will also facilitate grinding the seating surfaces.

Assembly or disassembly loads must never be applied in such a manner that the rolling elements are subjected to a load.

In designing a shaft, it should not be assumed that the bearings will have any tendency to stiffen it. Clamping the inner race to the shaft is so often encountered that standard locknuts and lock washers are available to accomplish this. They are used as shown in Fig. 13-12. A keyway is cut through the threads to receive a tongue on the lock washer. The nut has several slots which facilitate tightening. The lock washer has a number of ears, some of which will line up with the slots in the nut when it is tight and may be bent down to lock the assembly securely.

The machining to receive the outer race must be done in a manner to assure alignment with the axis of the shaft when it is installed. When a bearing is to be placed at each end of the shaft, this can be most readily accomplished if both seats can be bored from one end.

The curve of axial deflection versus axial load for a ball bearing that can carry both a radial and a thrust load rises rapidly with an increase in load, then becomes much flatter. A pair of bearings can be so arranged as to reduce this deflection. They are mounted so that one takes thrust in one direction, the other in the other direction. Then, by means of axial adjustment, a thrust load is induced in each bearing. They are now operating higher on the curve, where the deflection is less for a given load. As the amount of preload is relatively small, the effect on the bearing life is negligible. Preloading also considerably reduces the radial deflection of angular-contact bearings.

The duplex pair mentioned in Sec. 13-4 have the races ground with a predetermined offset, so that when the races are mounted flush the proper preload is produced. Double-row bearings are available with the preload built into the bearing.

13-11 NEEDLE BEARINGS

Needle bearings are similar to roller bearings in that the rolling elements are cylindrical. However, the rollers have a small diameter and great length relative to the bearing diameter. Another difference is that often there is no separator or cage. Needle bearings require smaller radial space and greater length than ball or roller bearings for the same application. They cannot carry a thrust load, and they have greater friction and a lower maximum speed limit than ball or roller bearings. They are available with both an inner and outer race, with only an outer race, with only an inner race, or without either an inner or an outer race. When neither race is used, a retainer is used where there would be difficulty in assembly. But sometimes even this is eliminated. When a race is omitted, the surface that takes its place must be hardened and ground. If the full load-carrying capacity of the bearing is to be developed, the hardness must be a minimum of Rockwell C58.

Continuously fed oil is the preferred form of lubrication. However, grease is often used. When grease is used, some provision should be made to replenish the supply of grease, for there is little space in the bearing to act as a reservoir. When high speed is involved, oil should be used.

13-12 SELF-ALIGNING BEARINGS

The maximum permissible misalignment of cylindrical roller and needle bearings is about 0.001 in./in.; for radial bearings the limit is several times this great; and for tapered roller bearings it is only one-half as large.

The purpose of a self-aligning bearing is to allow for inaccuracies in shaft alignment and deflections of the shaft or its supports that are greater than this. When deflections are provided for in this manner, stresses that otherwise would be induced are avoided.

Self-aligning ball bearings are classed as internal or external. The internal type has a spherical race upon which the balls roll and is often of the double-row type, as shown in Fig. 13-13a. Because the radius of the race is so much greater than the radius of the ball, less contact area is developed, and thus the load-carrying capacity is reduced. In the external

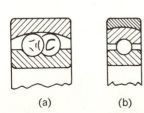

(a)   (b)

Fig. 13-13 Self-aligning ball bearings.

(a)   (b)

Fig. 13-14 Self-aligning roller bearings.

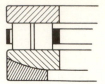

Fig. 13-15   A flat-race thrust bearing.   Fig. 13-16   A self-aligning roller thrust bearing.

type, the outer surface of the outer race is spherical, which fits in an added race, the inner surface of which is spherical, as shown in Fig. 13-13b. This type has essentially the same radial and thrust capacity as a single-row bearing of similar size.

Both single-row and double-row self-aligning roller bearings are available. There are several forms of each type. One form of each is shown in Fig. 13-14. Both are internally self-aligning; however, the load-carrying capacity is not affected.

13-13   BALL AND ROLLER THRUST BEARINGS

This section concerns bearings intended to carry only a thrust load.

The flat-race type shown in Fig. 13-15 develops a small contact area, and thus has low load-carrying capacity. The advantages are the low friction and the fact that the shaft running eccentrically has no effect on it. Ball thrust bearings are made with grooved races, also with two rows of balls.

Rollers are often used in thrust bearings. The cylindrical roller has the disadvantage that, because the outer end must travel farther than the inner end, considerable sliding takes place. This is reduced by making the rollers short. A tapered roller can have pure rolling. Both the ball and the roller type are available with a self-aligning feature, which compensates for slight misalignment. Any misalignment would cause only a few of the rolling elements to carry the entire load and thus shorten the life of the bearing. The outside face of one of the races is spherical and rests against a washer with a corresponding spherical surface, as shown in Fig. 13-16.

13-14   LINEAR-MOTION BEARINGS

Linear-motion bearings are available for use with both flat surfaces and shafts. Those for flat surfaces most often use rollers, and those for shafts use balls as the rolling element. The rolling element must circulate; that is, it must roll to the end of the bearing and then in some manner

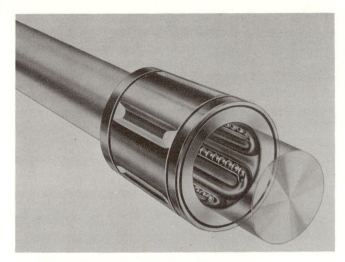

Fig. 13-17 A ball bushing. (*Thompson Industries, Inc.*)

return to the other end. The type for use with a shaft, known as a ball bushing, is shown in Fig. 13-17.

The surface in contact with the rolling element must be hardened and must have a smooth surface to develop the full capacity of the bearing. Linear-motion bearings, like all rolling bearings, are adversely affected by dirt. Because the surface upon which they roll is, in effect, a part of the bearing, preventing the accumulation of dirt or removing it before the bearing arrives is essential. Another item to be kept in mind when considering a ball bushing is that it is intended only for linear motion and should not be subjected to rotation.

13-15 PREMOUNTED BEARINGS

Premounted bearings are an assembly consisting of bearing, housing, provision for lubrication, and seals to retain the lubricant and exclude dirt. In addition, some permit altering the position of the bearing to provide for adjusting belts and chains. They are available in many forms and a wide range of sizes. Examples are shown in Figs. 13-18 and 13-19.

The bearing may be of the sliding, ball, or roller type. Some bearings incorporate a self-aligning bearing to allow for minor misalignment and deflection. Others are so designed as to allow for a difference in the thermal expansion of the shaft and the structure to which the bearing is attached.

The main advantages of premounted bearings are: they cost less than

Fig. 13-18 A premounted bearing incorporating a ball bearing. (*Fafnir Bearing Co.*)

units designed and made for each individual application, and they are readily available for replacement.

13-16 BEARING SELECTION

In the selection of a bearing for a particular application, the factors listed below deserve consideration. Some are more important in one application than another. Nevertheless, they are presented in the order in which it would be appropriate to consider them for the usual application.

1. Load — radial, thrust, steady, shock, magnitude
2. Life — number of revolutions required
3. Failure — probable type, consequences
4. Speed — constant, high, low, oscillating
5. Accuracy of shaft position under varying load
6. Tolerance to dirt
7. Tolerance to misalignment and deflection
8. Space required
9. Cost — first, replacement, designing
10. Power absorbed by friction
11. Lubrication requirements
12. Damping capacity
13. Availability

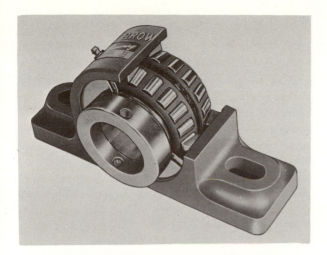

Fig. 13-19 A premounted bearing incorporating two tapered roller bearings. (*Browning Mfg. Co.*)

14. Maintenance required
15. Ease of replacement
16. Temperature
17. Corrosion

The advantages of rolling and sliding bearings are given below. They apply only in a general way and for the average situation.

Advantages of rolling bearings:
More precise shaft location under varying load
Low friction under moderate load
Ease of providing adequate lubrication
Superior when motion must be begun under load
Large momentary overloads can be carried
Less sensitive to interruption of oil flow
Shorter axial length
Many types in many sizes readily available
Manufacturers provide information to facilitate selection

Advantages of sliding bearings:
Less noise and vibration
May have infinite life
Require less space

Less sensitive to dirt
More suitable for shock load
Less damage from major bearing failure

QUESTIONS

13-1 Discuss contact area as applied to rolling bearings.

13-2 What are the functions of the separator?

13-3 What type of failure is unavoidable in rolling bearings?

13-4 What is the difference between a Conrad type and a filling-slot bearing?

13-5 In what manner is a Conrad type superior to a filling-slot type?

13-6 In what manner is a filling-slot type superior to a Conrad type?

13-7 What is the major difference between a Conrad type and an angular-contact type?

13-8 What is the major advantage of a double-row angular-contact type over a single-row angular-contact type?

13-9 What are the two basic kinds of radial roller bearings?

13-10 Comment on the thrust capability of a tapered roller radial bearing.

13-11 What effect does reducing the load to one-half have on the life of a ball bearing?

13-12 What is the difference between a shield and a seal?

13-13 Briefly discuss lubrication as applied to rolling bearings.

13-14 What is the advantage of installing the outer race in such a manner that a very slight rotational creep takes place?

13-15 What are the advantages of preloading?

13-16 What is the advantage of a needle bearing?

13-17 What is the difference between an internal and an external self-aligning bearing?

13-18 What are the advantages of a flat-race-type ball thrust bearing?

13-19 What is the disadvantage of a long cylindrical roller in a roller thrust bearing?

13-20 Are self-aligning thrust bearings available?

13-21 What is a ball bushing?

13-22 What is a premounted bearing?

13-23 List 10 factors that should be considered in selecting a bearing.

13-24 List several important advantages of rolling bearings.

13-25 List several important advantages of sliding bearings.

PROBLEMS

13-1 What is the life of a 10 series No. 10 bore bearing, carrying a radial load of 2000 lb?*

13-2 Select an 03 series bearing to carry a 2500-lb radial load and have a life of 2.5 million revolutions.

13-3 A unit is to have a life of 5 years, operating 8 hr a day 5 days a week. The radial load is 6000 lb, and the speed is 70 rpm. Select a series 73 bearing.*

13-4 Is a series N3 No. 12 bore bearing suitable for use at a speed of 1800 rpm?

13-5 A bearing is subjected to a radial load of 1300 lb and a thrust load of 2000 lb. What is the equivalent radial load?*

13-6 A bearing is subjected to a radial load of 3500 lb and a thrust load of 800 lb. What is the equivalent radial load?

13-7 Select a 10 series bearing to carry a radial load of 4200 lb and a thrust load of 3800 lb and have a life of 100 hours at 900 rpm.

13-8 What is the greatest radial load an 03 series No. 13 bore can carry if the inner race is stationary and the life must be 1 million revolutions?*

13-9 A radial roller bearing with a rib has a specific dynamic capacity of 14,300 lb. What is the maximum thrust load to which it should be subjected?*

13-10 Is an 01 series No. 18 bore bearing, turning 6000 rpm and carrying a load of 8000 lb, suitable for a life of 2.5 hours?

SELECTED REFERENCE

Wilcock, D. F., and E. R. Booser: "Bearing Design and Application," McGraw-Hill Book Company, New York, 1957.

14 BELTS AND CHAINS

14-1 INTRODUCTION

In mechanical design it is often necessary to provide for the transmission of power from one shaft to another that is parallel to it, or to make one shaft rotate faster or slower than another. This is usually accomplished by means of belts, chains, or gears. Gears will be discussed in the next chapter. Belts and chains are sometimes referred to as wrapping connectors, because they wrap around the pulleys or sprockets.

Factors in selecting a drive, that is, the means of connecting one shaft to another, are:

1. Center distance — The distance between the centers of the two shafts. For small distances, a gear is indicated, and for large distances, a chain or belt.

2. Timing — Causing one shaft to turn in a precise relation to the other. Belts slip and are, therefore, not suitable where timing is a factor. With chains, the proper number of revolutions is maintained, though the angular velocity is not constant. For circular gears, the timing is perfect.

3. Cost — The following drives are listed in the order of increasing cost: flat belts, V-belts, roller chains, silent chains, gears. Chains and gears require greater accuracy of alignment and a means of lubrication, which increases their installed cost.

4. Maintenance — Occasional adjustment is required for chains because of wear, and for belts because of stretch. The lubrication of chains and gears requires attention, and belts require dressing.

The design calculations for elements that depend on friction are subject to considerable uncertainty, because the value of the coefficient of friction is not known accurately for a particular situation and may change with the operating conditions.

14-2 THEORY OF WRAPPING CONNECTORS

As indicated in the last section, there are two types of wrapping connectors: those that depend on friction and those whose drive is positive. It should be pointed out that in some instances the slipping of belts is an advantage. In the event of jamming or stalling, a belt will slip and thus act as an overload protector to prevent damage to the machine.

The velocity ratio is the ratio of the angular velocity of the driving shaft to the angular velocity of the driven shaft. Thus

$$\frac{N_1}{N_2} = \frac{D_2}{D_1} \qquad (14\text{-}1)$$

and $N_1 D_1 = N_2 D_2$ $\qquad (14\text{-}2)$

where N_1 = rpm of driving shaft
N_2 = rpm of driven shaft
D_1 = diameter of driving pulley
D_2 = diameter of driven pulley

For a flat belt, the thickness of the belt is ignored, and the diameter of the pulley is the outside diameter.

For a chain drive, the following relation must be used

$$\frac{N_1}{N_2} = \frac{n_2}{n_1} \qquad (14\text{-}3)$$

where n_1 = number of teeth on driving sprocket
n_2 = number of teeth on driven sprocket

14-3 FLAT BELTS

Leather has had the most extensive use in the past and has characteristics that recommend it today. The most desirable portions of the hide may be cemented together to produce a belt of any desired thickness, width, and length. Leather has a high coefficient of friction, which tends to increase with proper use. Because of the ability of leather to stretch slightly without damage, and the tendency of friction to increase with slip, it has a higher momentary overload capacity than other belts.

Canvas belts are folded and stitched to make the required number of plies. They are usually impregnated with some substance to make them waterproof and to prevent damage to the fibers. They are cheap and are often used in hot, dry environments, or where they will receive little maintenance. Canvas belts are usually made in 4, 5, 6, and 8 ply.

A rubber belt is made by surrounding fabric with rubber, which is vulcanized to bond all the plies together. Cords are sometimes included to increase the strength. A belt may be made endless at the factory, or any of the standard belt fasteners may be used to join the ends. Rubber belts are used where exposure to moisture is required. They are adversely affected by sunlight and oil, and they deteriorate with time even when not in use. They are usually made in 3 to 10 ply.

Balata belting is made much like rubber belting, except that balata gum is used. It is not vulcanized. It does not deteriorate when not in use the way rubber does, but it should not be used where the belt temperature exceeds 120°F.

Woven belts are made by weaving on a loom to the desired thickness. They are appropriate where little power is to be transmitted at high speed. A comparatively recent devlopment is a composite belt consisting of a plastic core and leather facing. The plastic provides the strength and the leather provides a high coefficient of friction. This kind of belt is well suited to shock loads, but it costs more than a leather belt.

The velocity at which a belt travels may be found with

$$V = \frac{\pi DN}{12} = 0.262DN \tag{14-4}$$

where V = velocity, fpm
$\quad\quad D$ = pulley diameter, in.
$\quad\quad N$ = rpm

From Eq. (14-2), it is evident that it does not matter whether the driver or the driven pulley diameter is used; however, the diameter and rpm must be for the same pulley. An appropriate maximum velocity is 4000 fpm. Higher velocities are used, but the greater centrifugal force considerably reduces the effectiveness of the belt, and the life of the belt is also reduced.

The minimum pulley diameter for leather belts is: single ply — 3 in., double ply — 7 in., triple ply — 20 in. The minimum pulley diameter for rubber belts is: 3 ply — 4 in., 4 ply — 6 in., 5 ply — 8 in., 6 ply — 12 in., 7 ply — 16 in., 8 ply — 20 in., 9 ply — 24 in., 10 ply — 28 in. The angle of contact is the arc on the pulley that is in contact with the belt. For pulleys

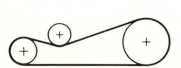

Fig. 14-1 Use of an idler pulley.

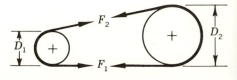

Fig. 14-2 Forces in a belt.

of equal diameter, it is 180° for each. For pulleys that are not of equal diameter, the smaller has less than 180° and the larger diameter more than 180°.

The angle of contact should not be less than 160°. To increase the angle of contact, an idler pulley may be used, as shown in Fig. 14-1. It is best if the idler pulley applies force by means of a weight rather than by adjustment or spring, and it should be placed near the small pulley on the slack side. An idler pulley should be avoided wherever possible. The upper belt should be the slack side.

Referring to Fig. 14-2, where F_1 is the tension in the tight side and F_2 is the tension in the slack side, it is evident that the torque at the small pulley is

$$T_1 = (F_1 - F_2)\frac{D_1}{2} \tag{14-5}$$

And the torque at the large pulley is

$$T_2 = (F_1 - F_2)\frac{D_2}{2}$$

The term $(F_1 - F_2)$ is known as the net tension.

In Sec. 11-3, the relation between torque and power was developed. When these concepts are combined with net tension and belt velocity, as found with Eq. (14-4), the equation for the horsepower transmitted by a belt becomes

$$\text{hp} = F_e \times \frac{0.262DNW}{33,000} = \frac{F_eDNW}{126,000} \tag{14-6}$$

where F_e = net tension, lb/in. of belt width
W = belt width, in.

The actual or effective net tension is dependent on the strength of the belt, the coefficient of friction between the belt and the pulley, the angle of contact, the initial tension in the belt, and the effect of centrifugal force on the belt. The catalog of the manufacturer of the belt to be used should be consulted for precise data. However, appropriate values for the net tension suitable for preliminary investigation are as follows. Leather belts: medium single-ply — 55 lb, medium double-ply — 95 lb, medium triple-ply — 130 lb. Rubber belts using light fabric: 13 times the number of plies. These values are for an angle of contact of 180°; 1 percent should be added for each 2° when the angle is more than 180°, and 1 percent should be subtracted for each 2° when the angle is less than 180°.

If a line connecting the centers of the pulleys makes an angle greater than 60° with the horizontal, the horsepower transmitted should be reduced 1 percent for each 1° over 60°.

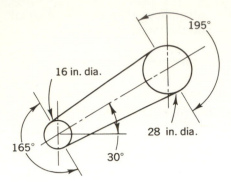

Fig. 14-3   Example 14-1.

Example 14-1   If a medium double-ply leather belt 12 in. wide is used in the situation shown in Fig. 14-3, how much power is transmitted?

Solution   The small pulley will transmit the least power, because of the smaller angle of contact. Thus, it should be used for the calculations.

The revolutions per minute of the small pulley is found with Eq. (14-2): $N_1 D_1 = N_2 D_2$, thus

$$N_1 = \frac{28 \times 600}{16} = 1050 \text{ rpm}$$

A medium double-ply leather belt has a net tension rating of 95 lb. The horsepower is found with Eq. (14-6)

$$\text{hp} = \frac{F_e D N W}{126,000}$$

$$\text{hp} = \frac{95 \times 16 \times 1050 \times 12}{126,000} = 151.5$$

This must be reduced by 7.5 percent to allow for the 165° angle of contact. Thus, the power transmitted is 140.2 hp. Because the angle that the line connecting the shaft centers makes with the horizontal is less than 60°, no allowance is necessary.

14-4 V-BELTS

V-belts consist of fabric and cords, usually made of synthetic fiber, molded in rubber in somewhat the same manner as flat rubber belts. They have a cross section as shown in Fig. 14-4. In some, steel cables are used in place of cords. They are quiet, clean, and well suited to short center distances. Because of the tendency to pull into the pulley groove under a heavy load, the effect of shocks is greatly reduced. They are about 2 percent less efficient than flat belts. An advantage of using several belts in a single drive is that the machine need not be immediately stopped when one belt fails.

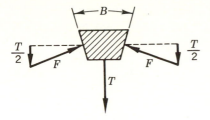

Fig. 14-4 Forces on a V-belt.

As shown in Fig. 14-4, the force upon which the friction is dependent is greater than the tension in the belt. The smaller the included angle *B*, the greater the effective coefficient of friction. However, the force that is required to pull the belt out of the groove also increases as the angle is made smaller. The best compromise is about 40°.

A V-belt drive may consist of a single belt or of several belts; when it consists of several belts, it is known as a multiple drive. When a multiple drive is used, the shafts must be parallel and the grooves in the pulleys must be identical, so that each belt will carry its portion of the load. And if one belt must be replaced, all the belts should be replaced. Otherwise, the new belt, which is not worn and has not stretched as much as the old belts, will be subjected to much more than its share of the load, and its life will be greatly shortened. A multiple drive is shown in Fig. 14-5.

Fig. 14-5 Multiple belt drive. (*Maurey Mfg. Corp.*)

As the belts are endless, the design must provide for using standard lengths. The design must also provide for the replacement of belts. The approximate length of a belt may be found with the equation

$$L = 1.57(D + d) + 2C + \frac{(D - d)^2}{4C} \tag{14-7}$$

where L = belt length, in.
$\quad D$ = diameter of large pulley, in.
$\quad d$ = diameter of small pulley, in.
$\quad C$ = distance between pulley centers, in.
For the larger belts, the pulley diameters are the pitch diameter, and the belt length is at its pitch line.

The center distance should not be smaller than that which provides a 120° angle of contact. The smaller angle of contact may be approximated with

$$\theta = 57.3\left(\pi - \frac{D - d}{C}\right) \tag{14-8}$$

where θ = angle of contact, degrees. Generally, the center distance is made not less than the diameter of the large pulley and not more than the sum of the diameters of both pulleys.

A V-belt is most efficient at about 4000 fpm. When Eqs. (14-1), (14-2), and (14-4) are used with V-belts, the pitch diameter of the pulley should be used.

A number of small belts, or one or a few large ones, may be used to transmit the same power. The smaller ones can use smaller pulleys, and

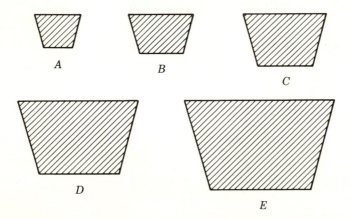

Fig. 14-6 Standard V-belt cross sections.

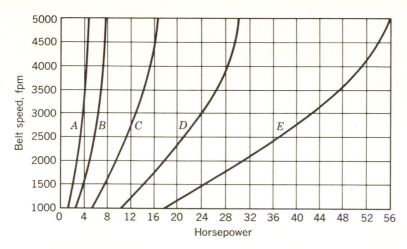

Fig. 14-7 Horsepower ratings of V-belts.

the center distance may be smaller, but the pulley width will be greater. The usual procedure is to select what is regarded as an appropriate belt size and then calculate the number of belts that are required to transmit the power. The sizes of standard V-belts are given in Fig. 14-6, and the horse-power rating is given in Fig. 14-7. These ratings are for the following pulley diameters: belt A — 5 in.; belt B — 7 in.; belt C — 12 in.; belt D — 17 in.; and belt E — 28 in. These are not the minimum diameter that may be used; however, with smaller-diameter pulleys the rating is reduced.

Judgment must be exercised in making allowance for peak loads, uncertainties, and so on, keeping in mind that variations in the load, caused either by that which is driven or by the driver, considerably reduce the life of the belt. The capacity of the belt should be from 1.3 times the normal power for light duty to twice the normal power for heavy duty. The allow-ance can often be provided for merely by selecting a whole number of belts; for example, use 3 belts when calculations show that 2.6 are required.

There is no orderly system of lengths for V-belts. The B size is available in the widest range of pitch lengths, from 36 in. to 300 in. The pitch lengths less than 100 in. for the B size are given in Table 14-1.

TABLE 14-1 Pitch Lengths of Selected V-Belts (in inches)

36.8	54.8	67.8	82.8
39.8	56.8	69.8	84.8
43.8	61.8	72.8	86.8
47.8	63.8	76.8	91.8
49.8	65.8	79.8	98.8
52.8			

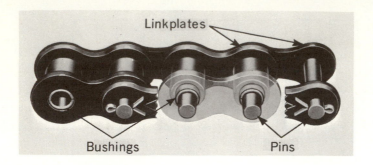

Fig. 14-8 Construction of a roller chain. (*Diamond Chain Co., one of the AMSTED Industries.*)

14-5 ROLLER CHAINS

A chain drive has some of the advantages of both a belt and a gear. It may be used for large or small center distances and is more compact than a belt drive. It can absorb a relatively greater shock load than a gear. It can operate at high or low temperature, and it has a high efficiency.

The two most often used forms of chain drive are the roller chain and the silent chain. The construction of a typical roller chain is shown in Fig. 14-8. Each part is designed to perform a single function best; no part is subjected to both stress and wear. A high-grade roller chain is made of hardened steel parts. The pitch is the distance between adjacent pins.

A chain must be used with a sprocket, and a sprocket does not have a constant radius. Therefore, the chain continually moves in a radial direction as it approaches and leaves the sprocket. This causes a fluctuation in the speed of the driven sprocket. The larger the number of teeth, the smaller this effect.

The average velocity of a chain is found with

$$V = \frac{pNn}{12} \tag{14-9}$$

where V = average velocity, fpm

p = chain pitch, in.

N = rpm of sprocket

n = number of teeth on sprocket

Equations (14-1) and (14-2) are applicable, providing the numbers of teeth on the sprockets are used in place of the diameters. The recommended maximum velocity is about 1500 fpm, though much higher velocities have been used successfully, especially with the smaller chains.

The best results are obtained when the center distance is about 40 times the pitch of the chain. Twice this is the maximum, and the minimum should not be less than that which will provide an angle of contact of 120°.

The recommended minimum number of teeth is 20, and the maximum is 120, though slow-speed drives may exceed these limits. If the velocity ratio is greater than 8:1, the speed change should be made in two steps, that is, with two separate sets of chain and sprockets. Sprockets should be hardened if they are small, if they run at high speed, or if they are heavily loaded.

The approximate length of a chain can be found with

$$L = \frac{n_1 + n_2}{2} + \frac{2C}{p} + \frac{p(n_1 + n_2)^2}{39.5C} \tag{14-10}$$

where L = length of chain in links
n_1 and n_2 = numbers of sprocket teeth
C = center distance, in.
p = chain pitch, in.

An even number of links should be used. Any whole number can be used, but when an odd number is required an offset link is necessary. The center distance should be adjustable to compensate for wear; however, an idler sprocket can be used on either the inside or the outside of the chain (the inside is next to the sprockets). It should be placed in the slack side. The number of teeth should not be less than that in the small sprocket, and they should be as carefully machined as those of the other sprockets. Idler sprockets should be used with large center distances.

From Sec. 11-3 and Eq. (14-9), the equation for the horsepower transmitted by a chain becomes

$$\text{hp} = \frac{F_a V}{33,000} \tag{14-11}$$

where F_a = allowable tension in chain, lb
V = velocity of chain, fpm

Neglecting the centrifugal force, which at the recommended velocity is negligible, the commonly used equation for the allowable tension in the chain is

$$F_a = \frac{2,600,000A}{V + 600} \tag{14-12}$$

where A = projected area of the pin joint, in.^{2}. For standard chain, all the dimensions are relative to the pitch, and the projected area of the pin joint is

$$A = 0.273p^2 \tag{14-13}$$

Standard pitch sizes are: $\frac{1}{4}$, $\frac{3}{8}$, $\frac{1}{2}$, $\frac{5}{8}$, $\frac{3}{4}$, 1, $1\frac{1}{4}$, $1\frac{1}{2}$, $1\frac{3}{4}$, 2, and $2\frac{1}{2}$ in.

The horsepower from Eq. (14-11) is for average conditions. For better-than-average conditions and light, intermittent service, 25 percent greater power would be appropriate. And for poor conditions, for heavy duty, or where a long life is required, one-half this power would prove more satisfactory.

Example 14-2 A roller chain with a ½-in. pitch is used with a driving sprocket that has 30 teeth and turns 400 rpm. How much power is transmitted?

Solution The velocity is found with Eq. (14-9):

$$V = \frac{pNn}{12} = \frac{0.5 \times 400 \times 30}{12} = 500 \text{ fpm}$$

The allowable tension is found with Eq. (14-12):

$$F_a = \frac{2{,}600{,}000A}{V + 600}$$

$$= \frac{2{,}600{,}000 \times 0.273 \times 0.5^2}{500 + 600} = 161 \text{ lb}$$

The power is found with Eq. (14-11):

$$\text{hp} = \frac{F_a V}{33{,}000} = \frac{161 \times 500}{33{,}000} = 2.45 \text{ hp}$$

A multiple-strand chain may be used to provide greater capacity with a given pitch. The parts are identical to a single chain except that the pin is longer. The power found by multiplying the value from Eq. (14-11) by the number of strands should be reduced by about 15 percent, because the load will not be perfectly distributed among the strands.

Except for special chains, good alignment of the shaft is important, and the sprockets should be in the same plane. Shaft alignment is especially important for chains operated at high speed or for wide multistrand chains.

14-6 SILENT CHAINS

Though a silent chain is not silent, it produces considerably less noise than a roller chain. There is no sliding between the chain and the sprocket, and for some types there is no sliding at the pin joint. Thus, the efficiency may be as high as 99 percent. A silent chain is suitable for high-speed operation, because of the smooth action. Silent chains are used to transmit from less than 1 hp to several thousand horsepower.

The construction of a representative silent chain is shown in Fig. 14-9. Some means must be provided to prevent the chain from slipping off the

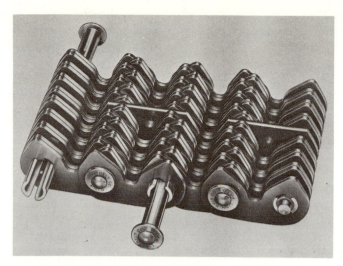

Fig. 14-9 Construction of a silent chain. (*Link-Belt.*)

sprocket. Some of the methods are: the outside links are made to fit over
the teeth; a special link in the center of the chain fits in a groove in the center
of the sprocket; or the sprockets are made with side plates.

The maximum velocity for usual applications is 2500 fpm. The life
of the chain is less at high velocity. The permissible velocity is less when
there is a small number of teeth on the small sprocket. For velocities less
than 1200 fpm, a roller chain is generally preferred. For a properly designed silent-chain drive, the velocity ratio may be as high as 12:1, though
making the speed change in two steps may be more economical for ratios
over 8:1. More than 150 teeth on the large sprocket, or fewer than 17 on
the small sprocket, should be avoided. An odd number of teeth on both is
desirable.

The standard pitches for one type of silent chain and the capacities
for the various sizes are given in Table 14-2. The horsepower transmitted
is

$$\text{hp} = \frac{T_p V W}{33,000} \tag{14-14}$$

where T_p = permissible tension, lb/in.

W = width of chain, in.

The total capacity of the chain should be greater than the normal power
to be transmitted. The amount is dictated by the application. For severe
shock, an appropriate chain capacity would be twice the normal power.

TABLE 14-2 Capacities of Silent Chain

PITCH, IN.	PERMISSIBLE TENSION, LB/IN.
$\frac{3}{8}$	75
$\frac{1}{2}$	100
$\frac{5}{8}$	125
$\frac{3}{4}$	150
1	205
$1\frac{1}{4}$	265
$1\frac{1}{2}$	335
2	600

Example 14-3 A silent chain of $\frac{3}{4}$-in. pitch is used to transmit 15 hp. The driving sprocket has 20 teeth and turns 350 rpm. What is the width of the chain?

Solution The velocity is found with Eq. (14-9):

$$V = \frac{pNn}{12} = \frac{0.75 \times 350 \times 20}{12} = 438 \text{ fpm}$$

The permissible tension from Table 14-2 is 150. Rearranging Eq. (14-14)

$$W = \frac{33,000 \text{ hp}}{T_p V} = \frac{33,000 \times 15}{150 \times 438} = 7.54 \text{ in.}$$

Lubrication becomes more important as the power transmitted by a chain or its speed increases. Providing a pool of oil that the chain dips into is a simple and effective method suitable for many applications. The best method is to spray oil on the inside of the lower span of the chain. If the quantity circulated is adequate, a desirable maximum temperature for the chain can be maintained. The viscosity of the oil must be low enough to permit its passing through the small clearances. One gallon per minute of SAE 20 oil is about right for an average installation. When the power is small and the speed is low, a chain need not be enclosed for those applications where it is not subject to dirt, and it may be manually lubricated with a brush or oil can.

14-7 OTHER WRAPPING CONNECTORS

There are a number of other wrapping connectors that deserve mention, though space does not permit details because they are suitable for use only in special situations.

Fig. 14-10   **A positive-drive belt.** (*Maurey Mfg. Corp.*)

Where considerable power is to be transmitted, and a very large center distance is involved, a rope drive may be used. A rope is used with appropriately grooved pulleys. This kind of drive was widely used in the past.

A flat link belt consists of leather links fastened together with pins in much the same manner as a roller chain. Such belts are more flexible than solid leather belts of equivalent thickness and are well suited to situations where considerable slippage may be encountered.

Round belts have a circular cross section and are made of leather or rubber. They are used in fractional horsepower drives, and are particularly suitable where the shafts are not parallel.

A double V-belt is available for situations that require the belt to fit a pulley first on one side, then on the other. Such a situation would occur where several pulleys were to be driven by the same belt.

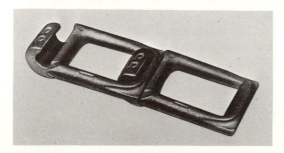

Fig. 14-11   **An open-link detachable chain.** (*Link-Belt.*)

Fig. 14-12 A pintle chain. (*Link-Belt.*)

A V-ribbed belt that is in effect a multiple-belt drive has been developed. It makes a more compact drive than the several V-belts that it replaces, and it eliminates the whipping that may take place in a multiple V-belt drive.

A link V-belt is a series of short links joined by pins to form a belt with a V cross section. These belts are easily made up to any length; also, assembly and adjustment for stretch is readily accomplished.

The connector shown in Fig. 14-10, called a positive-drive belt, is half belt and half chain, with advantages of both. A positive drive is obtained that may be used at a velocity as high as 16,000 fpm. When steel cables are used as the tension members, these belts can be used in fixed-center drives.

An example of an open-link detachable chain is shown in Fig. 14-11. Chains of this type are made from malleable-iron castings or pressed-steel links that are not machined. They are readily assembled to the desired length, but are suitable only for velocities less than 400 fpm. Where grit is a problem, the chain shown in Fig. 14-12, known as a pintle chain, is preferred, because the joint is enclosed.

A block chain is shown in Fig. 14-13. Block chains are cheaper than roller or silent chains. The velocity should be less than 900 fpm. The pitch is the distance between alternate pins.

A bead chain consists of small metal spheres joined by cylindrical pins of much smaller diameter — familiar as key chains. They are used for very

Fig. 14-13 A block chain. (*Diamond Chain Co., one of the AMSTED Industries.*)

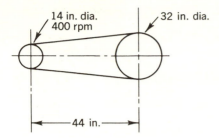

14 in. dia.
400 rpm

32 in. dia.

44 in.

Fig. 14-14   Prob. 14-5.

light loads and for velocities less than 200 fpm. Another kind suitable for very light loads is made up of links formed from wire; it is called a ladder chain.

QUESTIONS

14-1   Briefly discuss three factors that affect the selection of a drive.

14-2   What are the two basic types of wrapping connectors?

14-3   Define velocity ratio.

14-4   What are the advantages of leather as a belt material?

14-5   What are the advantages and disadvantages of rubber belts?

14-6   What is the purpose of an idler pulley?

14-7   What are the advantages of a V-belt?

14-8   In a multiple V-belt drive, why should all the belts be replaced when one fails?

14-9   What are the advantages of a chain drive?

14-10   Describe the construction of a roller chain.

14-11   What are the advantages of a silent chain compared with a roller chain?

14-12   Describe the construction of a silent chain.

14-13   Discuss the lubrication of chain drives.

14-14   What is a double V-belt, and how is it used?

14-15   Describe a positive-drive belt.

14-16   Compare an open-link detachable chain with a pintle chain.

PROBLEMS

Note: After solving some of these problems, it is *recommended* that the introduction to Chap. 19 be reread, Design Example 4 be studied, and Project 10 be completed.

14-1   In a flat belt drive, the pulley diameters are 8 and 20 in. The small one turns 800 rpm. How fast does the large one turn?*

14-2   The driving sprocket has 32 teeth and turns 650 rpm. The driven sprocket is to turn 300 rpm. How many teeth should be on the driven sprocket?

14-3   What is the velocity of the belt in Prob. 14-1?*

14-4 What is the angle of contact at the small pulley if the pulley diameters are 12 and 38 in. and the center distance is 4 ft?*

14-5 In Fig. 14-14, an 8-in.-wide 3-ply rubber belt is used. What power is transmitted?

14-6 What is the allowable tension in a roller chain of 1-in. pitch when the velocity is 1200 fpm?

14-7 What is the length of a V-belt if the pitch diameter of the pulleys is 18 and 32 in. and the center distance is 42 in.?*

14-8 A single-strand roller chain with a ⅝-in. pitch is used in a drive where the driven sprocket turns at one-third the speed of the driver and the driven sprocket turns 375 rpm and has 36 teeth. How much power is transmitted?

14-9 A chain drive uses a ½-in. pitch chain, the sprockets have 14 and 48 teeth, and the center distance is 28 in. What length of chain should be used?*

14-10 A silent chain of 1¼-in. pitch is used to transmit 32 hp. The driving sprocket has 24 teeth and turns 275 rpm. What is the width of the chain?

SELECTED REFERENCE

"Power Transmission Handbook & Directory," *Power Transmission Design Magazine*, Cleveland, 1964.

15

GEARS

15-1 INTRODUCTION

Gears can be used in situations where other means of transmitting power or causing a change in angular velocity would be undesirable or impossible. The requirements of a gear drive generally include the horsepower to be transmitted and the revolutions per minute of the two shafts. In some instances, the center distance is also specified. In addition to providing for these requirements, consideration should be given to the following:

1. Adequate life — Which may require that considerable attention be given to wear.
2. Alignment and deflections — Which involves the characteristics of the bearings, the deflection of the shafts and housing, and the accuracy of the machining.
3. Space — Gear drives can be so arranged as to require greatly different amounts and configurations of space.
4. Lubrication — Which has a great effect on life and efficiency.
5. Cost — Unless care is taken, a quality may be specified that is much higher than is necessary.

15-2 GEAR TOOTH ACTION

Points on the two cylinders in Fig. 15-1 have the same linear velocity if it is assumed that there is no slipping. It follows, therefore, that

$$\text{rpm of } A \times d = \text{rpm of } B \times D \qquad (15\text{-}1)$$

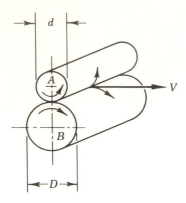

Fig. 15-1 Friction cylinders.

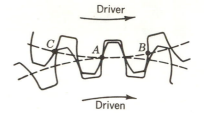

Fig. 15-2 Gear teeth in mesh.

To prevent slipping, teeth may be so arranged on each cylinder that the effective diameters are not changed, as shown in Fig. 15-2, where the dash lines are equivalent to the surfaces of the cylinders in Fig. 15-1. Adding the teeth does not change Eq. (15-1).

For the positive transmission of motion, the teeth need not be of any particular shape. But to avoid a great amount of noise and vibration, they must have a carefully determined and accurately reproduced shape. The most common shape is based on an involute curve. In Fig. 15-2 the teeth are in contact at point A. At point B, contact between the pair of teeth has just ceased, and at point C, contact is about to begin.

Fig. 15-3 A spur gear. (*Sier-Bath Gear Co., Inc.*)

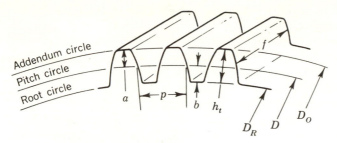

Fig. 15-4   Spur gear terminology.

15-3   SPUR GEARS

Spur gears are used to connect parallel shafts. The teeth of a straight-tooth spur gear are parallel to the shaft, as shown in Fig. 15-3. It is understood that this type is intended by the term spur gear. These gears can be used to transmit a large amount of power, and where noise is not objectionable, high velocities may be used. They impose only radial loads on their supporting bearings. Neither the center distance nor the axial location is critical.

Generally, two or more pairs of gears are used when the gear ratio is greater than 6:1. The shafts that turn at the higher speed can often be made smaller, because a smaller torque transmits the same power.

Spur gears are the easiest to manufacture and inspect, and are thus preferred unless the unique characteristics of some other type are required. Until otherwise specified, the discussion, though generally applicable to other types, will apply specifically to spur gears.

15-4   SPUR GEAR TERMINOLOGY

The following definitions and symbols are used in the discussion of gears. Some are illustrated in Fig. 15-4.

> **Pitch circle**   An imaginary circle that corresponds to the dash lines in Fig. 15-2.
>
> **Pitch diameter** (D)   The diameter of the pitch circle. The diameter of a gear is understood to be the pitch diameter.
>
> **Number of teeth** (n)   The number of teeth on the gear.
>
> **Diametral pitch** (P)   The number of teeth on the gear per inch of pitch diameter. $P = n/D$. The pitch of a gear is understood to be the diametral pitch.
>
> **Addendum** (a)   The radial distance from the pitch circle to the addendum circle. $a = 1/P$.

Dedendum (b)   The radial distance from the pitch circle to the bottom of the tooth space. $b = 1.157/P$.

Outside diameter (D_O)   The diameter of the addendum circle. $D_O = D + 2a$, also $D_O = (n + 2)/P$.

Root diameter (D_R)   The diameter of the root circle. $D_R = D - 2b$.

Whole depth (h_t)   The total height. $h_t = a + b$.

Face width (f)   The width of the gear tooth.

Circular pitch (p)   The distance measured along the pitch circle from a point on one tooth to the corresponding point on the adjacent tooth. $p = \pi D/n$. As $P = n/D$, $\pi = pP$.

Working depth (h_k)   The distance a tooth on one gear projects into the space between the teeth on the other gear. $h_k = 2a$.

Circular thickness (t_c)   The thickness of a tooth measured along the pitch circle. $t_c = p/2$.

All these dimensions are theoretical and are in inches. The relations apply to gears with either a 14½° or a 20° pressure angle. The pressure angle is the direction of the force that one tooth exerts on the other. It also determines the shape of the teeth in the following manner: In Fig. 15-5, the radial line passes through the center of both gears. The pitch circles of the gears are tangent at point Z. Through point Z a line is drawn perpendicular to the radial line; it is also tangent to both pitch circles. The pressure line is drawn through point Z at an angle with the tangent line equal to the pressure angle. The base circle, which is concentric with the pitch circle, is drawn tangent to the pressure line. The involute which forms the shape of the tooth is then generated using the base circle.

The size of a tooth is determined by the pitch, and the shape by the pressure angle. Therefore, if a gear of one diameter is to run with a gear of another diameter, the pitch and pressure angle must be the same for both, and of course, there must be a whole number of teeth on each.

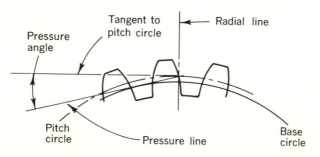

Fig. 15-5   Pressure angle.

Though gears can be made to any pitch, cutting tools are available for those that are more common. The recommended pitches from 4 to 24 are: 4, 5, 6, 8, 10, 12, 16, 20, 24. There are recommended pitches both larger and smaller than these. The 14½° pressure angle has been extensively used. However, 20° is generally preferred at present. The smaller gear in a set is often referred to as the pinion, and the larger gear as the gear.

15-5 POWER-VELOCITY-LOAD

Power is transmitted by one gear exerting a force on the other. This force is

$$F = \frac{33,000 \text{ hp}}{V} \tag{15-2}$$

where F = force on gear tooth, lb
   hp = horsepower transmitted
   V = velocity at pitch circle, fpm
The velocity at pitch circle is

$$V = 0.262DN \tag{15.3}$$

where N = rpm.

15-6 GEAR STRENGTH

The capacity of a gear to transmit power is dependent on the strength of a tooth acting as a cantilever beam. The allowable load is

$$F_s = \frac{sfY}{P} \tag{15-4}$$

where F_s = allowable load, lb
   s = allowable stress, psi
   Y = tooth form factor
The allowable stress may be obtained from Table 15-1, and the form factor for gears with a 20° pressure angle from Fig. 15-6.

The load found with Eq. (15-4) is the allowable load only for machined gears operated at a few revolutions per minute and without shock. Even if no shock is involved when one tooth transmits the load to another the load is suddenly applied. Slight inaccuracies in the teeth induce additional stress. Both of these conditions become more severe as the velocity increases. However, an accurately made gear is less affected than one that is less accurate. To allow for these conditions and the quality of the gear,

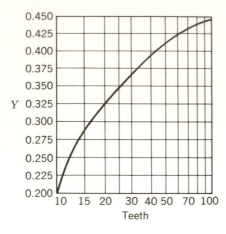

Y

Teeth

Fig. 15-6   Form factor.

the load found with Eq. (15-4) is multiplied by a factor K. The value of K for commercial-quality gears at velocities less than 2000 fpm is

$$K = \frac{600}{600 + V} \tag{15-5}$$

For accurately hobbed or generated gears at velocities less than 4000 fpm, the value is found with

$$K = \frac{1200}{1200 + V} \tag{15-6}$$

For precision gears that are ground or lapped and operating at velocities over 4000 fpm, K is found with

$$K = \frac{78}{78 + \sqrt{V}} \tag{15-7}$$

The force F_s found with Eq. (15-4) must exceed the force F found with

TABLE 15-1   Allowable Stress for Materials Used in Gears

MATERIAL	STRESS, PSI
Ordinary cast iron	6,000
Highest-grade cast iron	13,000
Untreated steel	20,000
Case-hardened steel	50,000
Heat-treated alloy steel	60,000
SAE 20 bronze	12,000
Rawhide	6,000
Bakelite	7,000

Eq. (15-2). Before comparing F and F_s, F should be increased to allow for any uncertainties and for the character of the load. In a situation involving little uncertainty and heavy shock, it would be appropriate to multiply F by 2. Where an extremely long life under continuous operation is required, gears should be designed on the basis of wear. This is not covered in this text.

When a pinion with a normal tooth form has less than a certain number of teeth, there will be an interference that will prevent proper meshing. This interference will take place at the base of the tooth on the pinion. If the pinion is made by a process that generates the teeth, such as shaping or hobbing, there will be no interference, because a normal tooth form is not produced. However, the thickness of the tooth at the base will be reduced. This is referred to as undercutting. It is undesirable because it weakens the tooth. The minimum number of teeth that involves neither interference nor undercutting for gears with a 14½° pressure angle is 32. For gears with a 20° pressure angle, the minimum is 18. By modifying the form of the tooth, the minimum number of teeth can be further reduced, and for gears with a 20° pressure angle and stub teeth, the minimum number is 14.

For a spur-gear drive to operate quietly, one gear must be made of nonmetallic material. Because of the higher cost of such material, it is usually used for the pinion. The velocity should not exceed 2500 fpm when a nonmetallic gear is used.

The face width of a gear is generally made from 3 to 4 times the circular pitch. For a gear made integral with the shaft, the root diameter must be slightly greater than the shaft diameter. For a gear that is keyed to the shaft, the minimum pitch diameter is twice the shaft diameter.

15-7 BACKLASH

It is the usual practice in making gear calculations to assume that the circular thickness of a tooth is equal to one-half the circular pitch. However, to prevent binding and to allow for slight inaccuracies and the effects of temperature, a clearance must be provided. This is accomplished by making the teeth on both gears, or on one gear, thinner than the theoretical thickness. This clearance is called backlash. For most applications, it is about $0.04/P$. The amount of backlash has no effect on the proper gear tooth action, though for applications where precise timing is required it must be very small.

15-8 GEAR TRAINS

A gear train is two or more gears that operate together. In Sec. 15-2, it was shown that the gear ratio of a pair of gears depends on their diameters. The number of teeth on each gear could also be used. Thus, a 30-tooth pinion in one revolution will cause a 60-tooth gear to make one-half a revolution.

The use of a gear train cannot increase power — actually there is a loss, which for most gear types is negligible. However, a gear train can, in accordance with Eq. (11-1), increase torque at the expense of rpm, or increase rpm at the expense of torque.

Example 15-1   A torque of 800 lb-in. is applied to a 6-pitch pinion which has 42 teeth. What is the torque produced by the 14-in. gear?

Solution   The torque at the gear will be that at the pinion multiplied by the gear ratio, which is the ratio of the diameters of the two gears or the ratio of the number of teeth on the two gears. $P = n/D$; therefore, a 6-pitch pinion with 42 teeth will have a diameter of 7 in., and the ratio is 2:1. Thus, the torque at the gear is $2 \times 800 = 1600$ lb-in. The alternate solution is to use $P = n/D$ to find the number of teeth in the gear, which is $6 \times 14 = 84$, and the ratio is 2:1.

When one gear drives another, the direction of rotation is reversed. If this is undesirable, the following arrangement may be used: gear *A* drives gear *B*, and gear *B* drives gear *C*. Gear *B* serves only to cause the desired direction of rotation; the gear ratio is unaffected. Gear *B* is called an idler gear.

15-9 INSTALLATION DESIGN

The shafts supporting the gears must be parallel, or the load instead of being distributed over the width of the tooth will be concentrated at one end. This may cause a tooth to break, and will at least cause excessive wear. An increase in center distance from that desired is less objectionable than a decrease.

With proper lubrication a thick-film condition tends to develop, which means that gears which are properly designed can have an extremely long life if clean oil of the proper viscosity is available in sufficient quantity. A gear should dip into the oil about 1 in. In most cases, the same oil may be used for both gears and bearings.

The bearings that support a shaft, which in turn maintains the proper location for a gear, must resist the force calculated with Eq. (15-2) and the force perpendicular to this force that tends to separate the gears, which is

$$S = F \tan \alpha \tag{15-8}$$

where S = separating force, lb
α = tooth pressure angle
If the gear is located an equal distance from each bearing, the load on each

bearing is one-half the resultant of the loads F and S. If the gear is not centered, the loads on each bearing may be found by the method used to find the reactions for a simple beam. It should be noted that the force F does not include any allowance for uncertainty or for the character of the load. This allowance should be made in accordance with the recommendations of Chaps. 12 or 13.

Example 15-2 A set of gears is located in the center of the shafts that are mounted on ball bearings. The gears have a pressure angle of 20° and transmit 25 hp. The pinion has a diameter of 4 in. and turns 900 rpm. What is the working load on each bearing?

Solution The force that one gear exerts on another is found with Eq. (15-2):

$$F = \frac{33{,}000 \text{ hp}}{V} = \frac{33{,}000 \times 25}{0.262 \times 4 \times 900} = 877 \text{ lb}$$

The force tending to separate the gears is found with Eq. (15-8)

$$S = F \tan \alpha = 877 \tan 20 = 319 \text{ lb}$$

The bearing load is the same for each bearing and is equal to one-half the resultant of F and S:

Resultant $= \sqrt{877^2 + 319^2} = 935 \text{ lb}$

Bearing load $= \dfrac{935}{2} = 468 \text{ lb}$

15-10 DIMENSIONS

A drawing of a gear should show all the details of the blank (the part before the gear is cut) with suitable tolerances. The gear teeth need not be drawn. The dimensions pertaining to teeth should include the diametral pitch, the pressure angle, and the number of teeth. It would be appropriate to consider backlash, and to specify the desired circular thickness of the tooth with a tolerance.

If both gears of a set are not to be made by the same manufacturer, information about the mating gear, such as material and number of teeth, should be specified. If the drawing is to be used in the inspection of the gear, additional dimensions may be desirable; however, superfluous interdependent theoretical dimensions should not be specified.

15-11 OTHER TYPES OF GEARING

There are several other types of gears that can be used to connect two parallel shafts. Their unique characteristics may recommend them for a specific application.

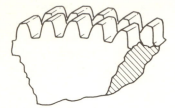

Fig. 15-7   A stepped gear.

A disadvantage of a straight spur gear is that the load is suddenly transferred from one tooth to the other. One way of avoiding this is to use a stepped gear. A gear with two steps would be, in effect, two spur gears of one-half the width, rotated so that the tooth of one was opposite the tooth space of the other and then permanently joined. This is shown in Fig. 15-7. A gear of this type is occasionally used where the teeth are cast to size.

A helical gear may be thought of as a stepped gear of an infinite number of steps. A helical gear is shown in Fig. 15-8. Because of the gradual engagement of the teeth, helical gears are quieter and stronger than spur gears, and may be operated at a higher velocity. The helix of one gear must be opposite that of the other; that is, a right-hand gear and a left-hand gear are run together. Because the teeth are not parallel to the shaft, a thrust load will be produced. The thrust will be in one direction for the right-hand gear and in the opposite direction for the left-hand gear if they revolve in only one direction.

To eliminate the thrust, one right-hand helical gear and one left-hand helical gear can be mounted on the same shaft. A single gear made with both a right-hand and a left-hand helix is called a herringbone gear. Be-

Fig. 15-8   A helical gear. (*Sier-Bath Gear Co., Inc.*)

Fig. 15-9 An internal gear. (*Sier-Bath Gear Co., Inc.*)

cause the thrust loads cancel for double helical gears and herringbone gears, the helix angle may be made greater. This improves the smoothness and strength compared with a single helical gear.

A set of herringbone gears requires precise axial alignment. Otherwise, one-half of the gear must carry the whole load.

An internal gear is one which has the teeth on the inside of the rim rather than the outside, as shown in Fig. 15-9. The direction of rotation is the same for both the internal gear and the mating pinion. Because of the concave-convex arrangement, there are more teeth in contact; thus, they run more quietly than a pair of external gears. An internal gear permits a much more compact arrangement; however, internal gears may be used only at the ends of parallel shafts. There should be at least 12 more teeth in the gear than on the pinion to prevent interference of the teeth.

A rack may be thought of as a gear of infinite diameter. When a gear with a fixed location and in mesh with a rack is turned, the rack will move in a straight line, or if the rack is moved, the gear will turn. Thus, either a gear or a rack can be the driver, and linear motion can be converted to rotation, or rotation to linear motion. The dimensions of a rack are as shown in Fig. 15-10. The pitch is equal to the circular pitch of the mating

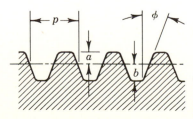

Fig. 15-10 Dimensions of a rack.

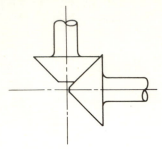

Fig. 15-11   Friction cones.

gear, the angle ϕ is the pressure angle, a and b are the addendum and dedendum of the gear. The smallest number of teeth on the gear that will not cause interference with the rack is 32 for a $14\frac{1}{2}°$ pressure angle and 18 for a 20° pressure angle.

15-12   BEVEL GEARING

A pair of spur gears was compared to two cylinders, one driving the other by friction. In a similar manner, bevel gears can be compared to two cones, as shown in Fig. 15-11. The more important dimensions of a bevel gear are illustrated in Fig. 15-12.

Pitch diameter (D)   The diameter of the base of the pitch cone.
Backing (x)   The distance from the base of the pitch cone to the rear face of the hub.

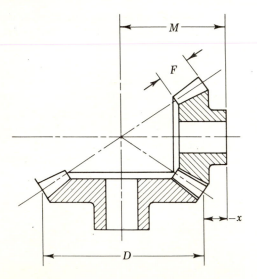

Fig. 15-12   Bevel gear dimensions.

Mounting distance (M)   The height of the pitch cone plus the backing.
Face (F)

The apex of both pitch cones must coincide at the intersection of the shaft center lines.

The pitch diameter is used to specify the size of a bevel gear. The gear ratio may be found by comparing the number of teeth on the gears. A point that must be kept in mind is that bevel gears are made in pairs and are not interchangeable in the same manner as spur gears.

The alignment of bevel gears is much more critical than spur gears. The shaft center lines must come very close to intersecting, and the axial location of the rear face of the hub must be very close to the mounting distance. The shaft and supporting members should be rigid. In addition to the radial load, the bearings will be subjected to the thrust developed in each shaft directed away from their point of intersection.

The gear ratio for bevel gearing is generally less than 4:1. When a pair of bevel gears have an equal number of teeth and the shafts are at 90°, they are referred to as miter gears. Bevel gears can be produced for shaft angles other than 90°.

A spiral bevel gear is similar to a straight bevel gear in much the same manner as a helical gear is similar to a spur gear. The advantages are: smoother tooth engagement, quieter operation, greater strength, and a higher permissible velocity.

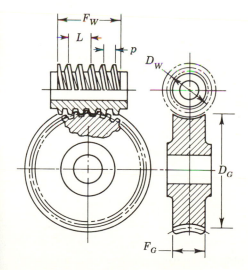

Fig. 15-13 Worm gear dimensions.

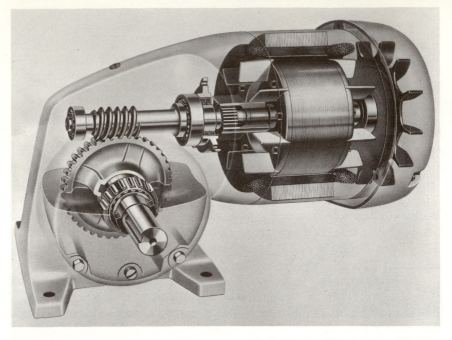

Fig. 15-14 An application of a worm gearset. (*U. S. Electrical Motors, Div. of Emerson Electric Co.*)

15-13 WORM GEARING

Worm gearing is used to connect nonintersecting shafts that are at right angles. It is extensively used for gear ratios greater than 5:1. When properly designed and lubricated, worm gearing is the smoothest and quietest form of gearing, though the efficiency is not as great as for most other forms.

A worm is similar to a screw and may have a single or multiple thread. The more important dimensions are illustrated in Fig. 15-13.

> **Pitch diameter of the gear** (D_G)
> **Pitch diameter of the worm** (D_W)
> **Axial pitch of the worm** (p)   The same as the pitch of a screw thread. The axial pitch of the worm and the circular pitch of the gear must be equal.
> **Lead of the worm** (L)   The same as the lead of a multiple-thread screw thread.
> **Face width of gear** (F_G)
> **Face length of worm** (F_W)

Very high gear ratios may be obtained with worm gearing; a ratio of 300:1 is not unusual. The ratio is equal to the number of teeth on the gear divided by the number of threads on the worm. Thus, a single-thread worm and a gear with 80 teeth has a ratio of 80:1.

Practically all worm gearsets that use a single-thread worm are self-locking; that is, the worm cannot be turned by turning the gear. This is because of the small helix angle of the gear and the friction involved.

Because sliding rather than rolling takes place between the worm and the gear, there is considerable friction. The capacity of the set is often limited by the rate of heat dissipation. To reduce the friction, the usual practice is to use a case-hardened alloy steel worm and a bronze gear. A very smooth finish on the worm and adequate lubrication are also essential.

A worm gearset must be accurately mounted, especially the axial location of the gear relative to the worm. The mounting must be rigid, and an effort should be made to keep the deflection of the worm shaft to a minimum. Provision must also be made for the thrust loads. Figure 15-14 illustrates an application of a worm gearset.

QUESTIONS

15-1 Do spur gears develop a thrust load?

15-2 Explain diametral pitch.

15-3 What is the addendum?

15-4 Explain circular pitch.

15-5 What is the pitch of a gear?

15-6 Explain pressure angle.

15-7 Why is a nonmetallic gear used?

15-8 What is backlash?

15-9 Why is backlash necessary?

15-10 What is the purpose of an idler gear?

15-11 What disadvantage of a spur gear does a stepped gear overcome?

15-12 Explain a stepped gear.

15-13 Explain a helical gear.

15-14 What are the advantages of a helical gear?

15-15 Explain a herringbone gear.

15-16 What are the important features of an internal gear?

15-17 What is a rack?

15-18 What is the function of a rack?

15-19 What is a miter gear?

15-20 What are four important characteristics of worm gearing?

PROBLEMS

Note: Upon completion of some of these problems, Design Example 5 and Project 11 or 12 in Chap. 19 are appropriate.

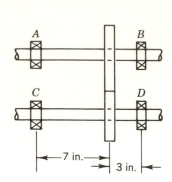

Fig. 15-15   Prob. 15-9.

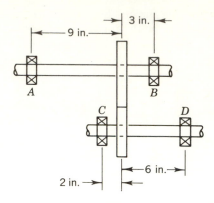

Fig. 15-16   Prob. 15-10.

15-1 For a 4-in. gear with 24 teeth, find the following: pitch, outside diameter, root diameter, circular pitch, and whole depth.*

15-2 What is the force that a 5-pitch 12-tooth pinion, turning 450 rpm, exerts on the gear in transmitting 4 hp?*

15-3 A 4-pitch 14-tooth pinion exerts a force of 600 lb on the gear in transmitting 7 hp. What is the revolutions per minute of the pinion?

15-4 What is the allowable load for a 5-in. 25-tooth gear, 2 in. wide, made of heat-treated alloy steel?

15-5 What is the gear ratio for a worm gearset in which the worm has a triple thread and the gear has 36 teeth?*

15-6 The torque is to be increased from 600 lb-in. to 1680 lb-in. with 12-pitch gears. The pinion is to have a diameter of 1.25 in. How many teeth must the gear have?

15-7 The speed is to be reduced from 1750 rpm to about 800 rpm with 16-pitch gears. The driver is to have 48 teeth. What is the diameter of the driven gear?

15-8 What is the force tending to separate two gears transmitting 5 hp, if the pressure angle is 20° and the small gear has a diameter of 3 in. and turns 1750 rpm?*

15-9 A set of gears is located as shown in Fig. 15-15. The gears have a pressure angle of 20° and transmit 56 hp. The gear has a diameter of 18 in. and turns 260 rpm. What is the load on each bearing?

15-10 A set of gears is located as shown in Fig. 15-16. The gears have a pressure angle of 14½° and transmit 150 hp. The gear has a diameter of 24 in., and the pinion turns 200 rpm. The gear ratio is 3:1. What is the load on each of the bearings?

SELECTED REFERENCE

Dudley, D. W.: "Practical Gear Design," McGraw-Hill Book Company, New York, 1954.

16 CLUTCHES AND BRAKES

16-1 INTRODUCTION

The primary purpose of a clutch is to connect or disconnect two shafts quickly, and for a brake it is to stop the rotation of a shaft. Though the operation of a clutch and a brake are essentially the same, there are important differences in the two units. The major difference is caused by the fact that in the case of the clutch, initially one part is rotating and one is stationary, but finally both are rotating. Whereas in the case of the brake, initially one part is rotating and one is stationary, but finally both are stationary. This difference greatly affects the design. Centrifugal force and the dissipation of heat are also factors that may have a considerable effect on the design.

16-2 FUNCTIONS OF A CLUTCH

A clutch may be required to connect a rapidly turning shaft to one that is stationary, or to cause the two to turn at the same speed and to do so in a manner such that shock is not produced. Or it may be used to limit the torque that is transmitted, or to prevent transmitting a torque in the reverse direction. A clutch may be required to transmit in a short time the large amount of energy that has been stored in a flywheel by a small power source over a long period.

Clutches vary in the means employed to transmit the torque, in the manner in which the parts of the clutch are attached to the shafts, and in the means of engaging and disengaging. There are also several ways of controlling a clutch: automatic control inherent in

Fig. 16-1   A square-jaw clutch. (*Link-Belt.*)

the design of a clutch, automatic control by a means external to the clutch, and manual control.

16-3 POSITIVE CLUTCHES

The most common type of positive clutch is the jaw clutch. There are two types: the square-jaw clutch, shown in Fig. 16-1, and the spiral-jaw clutch, shown in Fig. 16-2. Theoretically, an axial force is not required for proper operation, but this should not be relied on. One of the jaws must be on a feather key or a spline to permit the required axial motion.

The square-jaw type can transmit torque in either direction, but should not be used when engagement must be accomplished while a shaft is turning. The spiral-jaw type is easier to engage and should be used whenever engagement is frequent. It may be used when one of the shafts is turning, but a severe shock will be produced, and it should not be used if the shaft is turning more than 50 rpm.

The design of a jaw clutch is so similar to other elements that a detailed discussion is not necessary. Sufficient bearing area must be provided

Fig. 16-2   A spiral-jaw clutch. (*Link-Belt.*)

for the faces of the jaws, and adequate shear area must be allowed for in selecting the outside diameter and the diameter of the bore. The moment arm found by dividing the sum of the two diameters by 4 is reasonable.

16-4 PLATE CLUTCHES

A plate clutch is the most frequently used type and, therefore, will be discussed in detail. It depends on friction to transmit the torque and thus may slip. This ability to slip permits connecting a rapidly turning shaft to one that is stationary without producing shock. Also, the torque that is transmitted may be limited. When a clutch is slipping, heat is generated. If the clutch is used to cause two shafts to rotate at the same speed, heat is generated for only a short time, and if operation is infrequent, there will be time for it to cool. However, if frequent operation or continuous slipping is required, the clutch must be designed to continuously dissipate the heat generated.

A simple plate clutch is shown in Fig. 16-3. Part A is secured to its shaft, part B is keyed to its shaft with a feather key, the facing C is attached to B. Part B is forced against A to engage the clutch and away from A to disengage. D and d are the two diameters of the facing. Because the facing close to the axis of rotation is ineffective, d is seldom less than one-half of D.

The design of a clutch is based on the assumption that either the pressure is uniformly distributed or the axial wear is uniform, neither of which is actually the case. However, an equation that has proved successful is

$$T = \frac{fP(D + d)}{4} \qquad (16\text{-}1)$$

where T = torque transmitted, lb-in.
$\quad P$ = axial load, lb
$\quad f$ = coefficient of friction

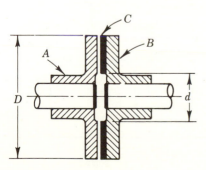

Fig. 16-3 A plate clutch.

The coefficient of friction depends on the facing material, the material of the part it rubs against, whether the surfaces are dry or oily, the temperature of the facing, and the pressure on the facing. A worn facing has a lower coefficient, and if there is considerable slipping, the coefficient is greatly reduced. In designing a clutch, the temperature and pressure are usually ignored and the coefficient for a worn facing is used. Values for the coefficient of friction and an appropriate allowable pressure are given in Table 16-1.

The inertia of the driven parts must be overcome when the clutch is engaged. The amount of slipping is dependent on the magnitude of this inertia. To allow for overcoming inertia and for the reduced coefficient of friction when there is slipping, the normal torque is multiplied by 1.5 to 2. In addition to this, a service factor should be included which is dependent on the character of the load. It varies from 1 for a steady load to about 3 for a load involving severe shock.

When the operation of the clutch is frequent and a long life is desired, it would be well to keep in mind that an increase in facing life accompanies a reduction in the pressure.

The capacity can be increased by increasing the number of friction plates; that is, adding an additional facing and surface for it to rub against doubles the capacity. Such a unit is known as a multiple-disk clutch. The capacity can be made several times greater in this manner, but there is a practical limit. Because it is such a compact unit, heating can easily become a problem. Also, the difficulty of obtaining complete disengagement increases with an increase in the number of disks.

A multiple-disk clutch is shown in Fig. 16-4. This type incorporates the means of engaging the clutch. When air is forced into the air tube A, it expands, forcing the two floating plates B and the two friction elements C against the backplate D, which is part of the hub. Thus, four friction

TABLE 16-1 Clutch Design Values

MATERIALS	COEFFICIENT DRY	COEFFICIENT OILY	ALLOWABLE PRESSURE, PSI
Cast iron on cast iron	0.20	0.07	150
Cast iron on steel	0.30	0.10	150
Cast iron on leather	0.50	0.15	10
Cast iron on asbestos fabric	0.40	0.25	40

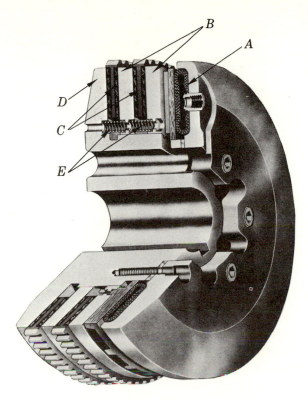

Fig. 16-4   A multiple-disk clutch. (*Wichita Clutch Co., Inc.*)

faces become effective in transmitting the torque, which is in direct proportion to the air pressure in the tube. When the air pressure is removed, the springs E (there are several sets) center the plates.

16-5   CONE CLUTCHES

The cone clutch shown in Fig. 16-5 is a simple friction clutch which has the advantage that a small axial load will produce a large force pressing the friction surfaces together. The angle α should not be less than 8° and is usually not more than 15°. The torque transmitted can be found with Eq. (16-1) when modified to allow for the angle as follows:

$$T = \frac{fP(D + d)}{4 \sin \alpha} \tag{16-2}$$

Though this type of clutch is not used as extensively now as it was in the past, it can often be used to advantage in a situation where a facing is undesirable and oil is present.

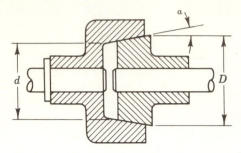

Fig. 16-5   A cone clutch.

16-6   OVERRUNNING CLUTCHES

It is often necessary to provide a means of transmitting power when the driver is rotating in one direction but which will automatically release if the direction of rotation is reversed or if the driven shaft begins to rotate faster. An overrunning clutch may be used in such a situation. It consists of a number of balls or cylindrical rollers in appropriately shaped spaces, as shown in Fig. 16-6. When the driver (inner member) rotates clockwise, friction causes the rollers to wedge between the driver and the outer ring, and power is thus transmitted. When the driver rotates counterclockwise, the rollers are forced in the opposite direction and wedging does not take place; thus, no power can be transmitted. The latter condition will also exist if the outer ring begins to rotate faster than the driver.

16-7   OTHER TYPES OF CLUTCHES

The force required to operate the clutches that have been discussed may be supplied by an electric, a hydraulic, or a pneumatic actuator. However, there are a number of clutches that employ a different means of transmitting the torque or of generating the necessary force.

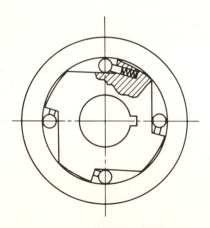

Fig. 16-6   An overrunning clutch.

There are several types of electric clutches that employ magnetic forces to couple the two halves of the clutch. There is what is sometimes called a dry-fluid coupling, which consists of a housing attached to the driving shaft and a rotor attached to the driven shaft. A quantity of steel shot is placed in the housing, which when the housing is rotated is thrown by centrifugal force to the outer portion of the housing where it packs against the rotor. With the rotor thus locked to the housing, the two shafts turn together. A true fluid coupling uses a liquid that is thrown by centrifugal force toward the outer portion of the driving half of the unit. This liquid strikes blades in the driven half, causing it to rotate.

There are several types of centrifugal clutches. One uses blocks of friction material in pockets so arranged in the driving half that they can move only in a radial direction. When the driving half is rotated, the blocks are thrown by centrifugal force against the driven half. The friction of the blocks rubbing against the driven half transmits the torque.

It is to be noted that all the clutches mentioned in this section, except the magnetic, are self-actuating, and the driven half is accelerated smoothly. They will slip under an overload. Thus, there is little tendency to transmit shock.

16-8 FUNCTIONS OF A BRAKE

It was mentioned earlier that the primary function of a brake is to stop the rotation of a shaft. A brake may also be used to control the speed of a shaft, or to prevent rotation when it is not desired. It may be part of a machine where the inertia of the moving parts must be resisted, or it may be attached to a drum that is involved in lowering a heavy weight, as in an elevator or crane, in which case the magnitude of the weight is the determining factor. There is also the familiar use in vehicles. As was the case with clutches, the operation of a brake should not generate a shock load in the associated mechanism.

16-9 BLOCK BRAKES

A block brake in its simplest form is merely a shaped block pressed against the shaft, as shown in Fig. 16-7. Usually the block is faced with a replaceable friction material, and it is forced against a drum which is attached to the shaft rather than directly against the shaft.

The force transmitted to the block by friction, known as the friction force, is

$$F_f = fF_n \qquad\qquad (16\text{-}3)$$

where F_f = friction force, lb
f = coefficient of friction
F_n = load forcing the block against the drum, lb

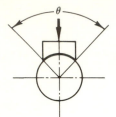

Fig. 16-7   A block brake.

This force may also be thought of as the resistance the block offers to the rotation of the drum. Thus, the resisting torque is

$$T = F_f r \qquad \text{or} \qquad T = f F_n r \qquad\qquad (16\text{-}4)$$

where T = resisting torque, lb-in.

r = radius of drum, in.

Equation (16-4) may be used for blocks of a length such that the angle of contact (θ in Fig. 16-7) is less than 60°. However, for larger angles, the assumption that the pressure of a block on the drum is evenly distributed is no longer valid and the variation must be taken into account. An equation that takes into account the variation in pressure is

$$T = \frac{4 f F_n r \sin (\theta/2)}{\theta + \sin \theta} \qquad\qquad (16\text{-}5)$$

where θ = angle of contact, rad.

Example 16-1   What is the resisting torque produced by a block brake faced with asbestos fabric pressed on a dry cast-iron drum with a force of 150 lb, if the angle of contact is 70° and the diameter of the drum is 18 in.?

Solution   The coefficient of friction from Table 16-1 is 0.40. The angle of contact in radians is $70/57.3 = 1.22$. The torque is found with Eq. (16-5):

$$T = \frac{4 f F_n r \sin (\theta/2)}{\theta + \sin \theta}$$

$$T = \frac{4 \times 0.40 \times 150 \times 9 \times 0.574}{1.22 + 0.940} = 573 \text{ lb-in.}$$

The force F_n is usually applied by means of a lever, as shown in Fig. 16-8. The location of the pivot point Z is important. Assuming the block and lever assembly in Fig. 16-8a to be a free body and taking moments about point Z, the moment of the friction force $F_f c$ is added to the moment Pd. If the pivot point is at Z in Fig. 16-8b, there is no moment for F_f. For the position shown by the dotted lines, the moment will be $F_f e$, and it will oppose the moment Pd and thus tend to release the brake. If the direction

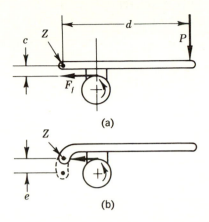

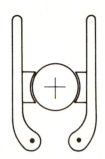

Fig. 16-8   Block brake arrangements.   **Fig. 16-9   A double-block brake.**

of rotation is reversed, the force F_f will be to the right and its moments will be reversed.

The arrangement shown in Fig. 16-8*a* requires less force to develop a given braking torque; however, it can result in the brake's grabbing, which is unsatisfactory or even dangerous in most applications.

A block brake imposes a load on the shaft and bearings that support the drum, which can be eliminated when two brakes are arranged as shown in Fig. 16-9.

16-10   BAND BRAKES

A band brake consists of a steel band lined on the inside with a friction material, as shown in Fig. 16-10. The operating force is applied so as to tighten the band around the drum. The difference in the tensions F_1 and

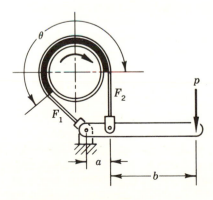

Fig. 16-10   A band brake.

F_2 is the friction force in the same manner as the difference in tensions was the driving force of a belt. The braking torque then is

$$T = r(F_1 - F_2) \tag{16-6}$$

where r = drum radius, in. The tensions can be found with

$$\frac{F_1}{F_2} = e^{\theta f} \tag{16-7}$$

and $F_1 = rbp_a$ (16-8)

where e = mathematical constant 2.718
θ = angle of contact, rad
b = band width, in.
p_a = maximum allowable pressure, psi

Example 16-2 A band brake faced with leather, which presses on an oily cast-iron drum, has the following dimensions: angle of contact — 210°, band width — 5 in., and drum diameter — 2 ft. What is the braking torque?

Solution The angle of contact is 3.66 radians. From Table 16-1 the coefficient of friction is 0.15 and the maximum allowable pressure is 10 psi.

$$e^{\theta f} = 2.718^{3.66 \times 0.15} = 1.73$$

F_1 is found with Eq. (16-8):

$$F_1 = rbp_a = 12 \times 5 \times 10 = 600 \text{ lb}$$

F_2 is found by rearranging Eq. (16-7):

$$F_2 = \frac{F_1}{e^{\theta f}} = \frac{600}{1.73} = 347 \text{ lb}$$

The braking torque is found with Eq. (16-6):

$$T = r(F_1 - F_2) = 12(600 - 347) = 3040 \text{ lb-in.}$$

The ends of the band can be connected to the arm in a manner to cause the tension in one end to aid in applying the brake. However, the relation between the direction of rotation and the band end that is connected to the fulcrum should be as shown in Fig. 16-9, if the drum rotates in both directions, or if it is desired to prevent self-locking. The force required to apply the brake is

$$P = F_2 \frac{a}{a + b} \tag{16-9}$$

16-11 BRAKE DESIGN

A brake is designed to produce a specified braking torque, but it is also involved in transforming energy into heat. Heating must be given consideration, for not only may the lining material be damaged by excessive heat, but the coefficient of friction is reduced for most linings as the temperature increases.

When the pressure on the friction surfaces is reduced, the heat generated per square inch is reduced and the size of the heat-dissipating surface is increased. The heat generated in a unit of time is the frictional force times velocity. The frequency of application determines the cooling that may take place. The environment also affects the cooling. The allowable pressure and coefficient of friction given in Table 16-1 may be used for brakes. The factor pV is also used as a guide. For intermittent application, relatively long cooling periods, and poor heat dissipation, the pV factor should be less than 55,000. If the load is continuously applied and the heat dissipation is poor, the pV should not exceed 30,000. For continuous application and good heat dissipation, such as is achieved with an efficient means of cooling, the pV may be as high as 85,000. For block brakes, the pressure is based on the projected area, that is, the length times the width. For band brakes, the area is the portion of the circumference in contact times the width.

If the actuating force is to be provided by some means such as air pressure, the design should provide for the failure of the source of actuating power. In most instances, the brake should be applied. This may be accomplished by designing the brake in a manner such that a spring causes it to be engaged, and the actuator applies the force to disengage it.

16-12 OTHER TYPES OF BRAKES

A single-disk brake is so arranged that when the brake is applied, the disk, which is the rotating part, is clamped between two blocks. The blocks cover a small portion of the disk and the remainder is free to dissipate heat.

A multiple-disk brake is similar to a multiple-disk clutch. In fact, some units in hoisting equipment are used as clutches when raising the weight and as brakes when lowering it.

Block brakes can be arranged so that the blocks press against the inside of the drum. In this case, the blocks are called shoes. The mechanism to force the shoes against the drum is a little more complex, and self-locking is accomplished more readily than with the external arrangement.

A hydrodynamic brake is essentially a fluid coupling, one half of which is stationary. The degree of braking action is controlled by regulating the amount of liquid in the brake. Because the liquid can be cooled by circulating it through a radiator, a great deal of energy can be absorbed without danger of overheating. However, this type of brake cannot stop or prevent rotation.

QUESTIONS

16-1 What are the functions of a clutch?

16-2 Compare the advantages and disadvantages of square-jaw and spiral-jaw clutches.

16-3 Describe the operation of a plate clutch.

16-4 Briefly discuss slipping.

16-5 Describe the operation of an overrunning clutch that uses cylindrical rollers.

16-6 Briefly discuss three types of clutches other than the positive and the plate types.

16-7 What are the two main functions of a brake?

16-8 Describe a block brake.

16-9 What is the major disadvantage of a block brake?

16-10 Describe a band brake.

16-11 What is the major advantage of a hydrodynamic brake?

16-12 What is an important limitation of a hydrodynamic brake?

PROBLEMS

Note: After solving some of these problems, completion of Project 13 in Chap. 19 is *recommended*, unless Project 15 is to be undertaken.

16-1 A plate clutch made of cast iron has a leather facing and is to operate dry. The outside diameter of the facing is 8 in., the inside diameter 4 in. The axial load forcing the facing against the plate is 300 lb. What torque is transmitted?*

16-2 A plate clutch has one cast-iron plate and one steel plate. The diameters are 10 and 6 in. How much power is transmitted at the maximum allowable pressure with oily surfaces, if the speed is 750 rpm?

16-3 A cast-iron cone clutch operates dry at 800 rpm. The diameters are 6 and 4 in., and the angle is 10°. What power is transmitted with an axial load of 125 lb?*

16-4 A cast-iron cone clutch operates dry at 600 rpm. The diameters are 8 and 5 in., and the angle is 12°. What axial load is required to transmit 4 hp?

16-5 A block brake with a short cast-iron block presses against an oily steel drum 3 ft in diameter with a force of 800 lb. What resisting torque is produced?*

16-6 A block brake with a short cast-iron block presses against an oily 3-ft-diameter steel drum that turns 800 rpm. What force is required to absorb 6 hp?

16-7 A leather-faced block brake is pressed against a dry 20-in.-diameter cast-iron drum with a force of 200 lb. The angle of contact is 100°. What is the resisting torque?

16-8 In Prob. 16-7, what is an appropriate width for the block?

16-9 A band brake faced with asbestos fabric presses against a dry cast-iron drum which is 4 ft in diameter. The angle of contact is 240° and the band is 10 in. wide. What is the braking torque?

16-10 The braking torque to be provided by a band brake is 2000 lb-in. The drum is dry cast iron, the facing is asbestos fabric, the angle of contact is 260°, and the band is 8 in. wide. What is the diameter of the drum?

SELECTED REFERENCE

"Power Transmission Handbook & Directory," *Power Transmission Design Magazine*, Cleveland, 1964.

17 SPRINGS

17-1 INTRODUCTION

Deformation is undesirable for most of the components dealt with in mechanical design, but a spring is specifically designed to deform under load. Springs are used to accomplish many different purposes where this ability to considerably deflect under load is desirable, such as to control loads due to impact, to exert a force, to control vibration, to store energy, and to measure forces. Springs perform vital functions in such diverse items as internal combustion engines, light switches, and typewriters. The detailed variations in springs are almost as numerous as their individual applications.

Springs are generally produced by organizations that specialize in their design and manufacture. It is good practice to make use of the extensive knowledge and experience of these specialists. This is done by providing them with a statement of the performance required and the various limitations. It is obvious that performance and limitations could be specified that would be difficult if not impossible to achieve, also that a performance such as exerting a given force at a given length could be accomplished by many different combinations of materials and spring sizes. Thus, it is essential that a designer who is engaged in the solution of a problem where a spring is involved determine the basic configuration while the design as a whole is still in the preliminary stage and specify performance and limitations that can be readily achieved.

Helical compression and extension springs account for the majority of springs made; therefore, these will be dealt with in detail in this chapter.

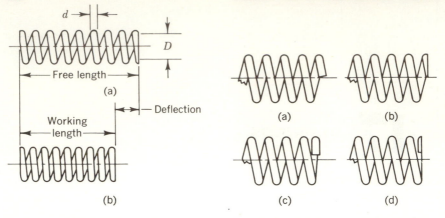

Fig. 17-1 Compression spring termi-   Fig. 17-2 Compression spring ends.
nology.

17-2 HELICAL COMPRESSION SPRINGS

Figure 17-1 shows a helical compression spring and illustrates the terminology. The type of end has an important effect on the performance of the spring and on how closely the performance will agree with that predicted by the equations used in its design. In developing the equations, the assumption is made that the load is axial. In the case of a spring with plain ends, as shown in Fig. 17-2a, the load is not axial; thus, this type of end should be avoided. Plain ends that are ground, as shown in Fig. 17-2b, are an improvement, but they should also be avoided when possible, because in the process of manufacture they tend to become a tangled mass. Ends that are closed or squared, as shown in Fig. 17-2c, do not tangle as much as the plain end. They are easy to manufacture and the load is more nearly axial. Springs made from wire smaller than 0.0317 with a diameter larger than 10 times the wire diameter should be made with this type of end unless the performance of the spring is critical. Ends that are squared and ground, as shown in Fig. 17-2d, ensure that the load will be axial; they should be used for all springs with wire diameter over 0.0317 where good spring performance is desired. This type of end is more expensive, of course, due to the grinding operation. If the ends are finished in any manner, part of the last coil becomes inactive, that is, it does not deflect the same as a normal coil; therefore, a deduction must be made from the total number of coils to allow for this in determining the number of active coils. If the ends are squared, or squared and ground, an allowance of one coil is made for each end; thus, the total number of coils is $N + 2$, where N is the number of active coils.

17-3 LOAD AND STRESS

The principal stress in a helical spring is torsion. A careful examination of Fig. 17-3 will make this clear. Figure 17-3a is a wire subjected to the torque Pr; Fig. 17-3b shows the wire partially formed into a coil; Fig. 17-3c is two views of the wire formed into a helical spring of 1½ coils. Other stresses will be discussed later. In the study of strength of materials it was shown that the shear stress induced in a rod by torsion can be determined by the equation $s_s = Tr/J$, where T is the torque in inch-pounds, r is the radius of the rod in inches, s_s is the shear stress in psi, and J is the polar moment of inertia. For a solid round rod, $J = \pi d^4/32$, where d is the diameter of the rod in inches. In the present problem, the moment arm is one-half the mean diameter of the spring. Thus, $T = PD/2$. Substituting in the original equation,

$$s_s = \frac{PD/2 \times d/2}{\pi d^4/32} = \frac{8PD}{\pi d^3} \qquad (17\text{-}1)$$

where D = mean diameter of the spring, in.

P = load on the spring, lb

In a similar manner, the equation for deflection of a rod subject to torsion can be used to produce the equation for the deflection of a spring:

$$F = \frac{8PD^3N}{Gd^4} \qquad (17\text{-}2)$$

where F = total deflection, in.

N = number of effective or active coils

G = modulus of elasticity in shear

As mentioned earlier, stresses other than shear due to torsion are

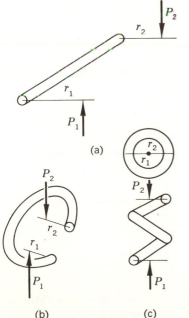

(a)

(b)   (c)   **Fig. 17-3 Torsion stress in a spring.**

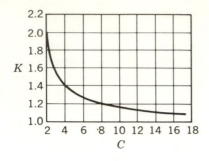

Fig. 17-4 Curve for Wahl factor.

induced in the spring. To allow for these additional stresses, a factor developed by A. M. Wahl and known as the Wahl factor is used. This factor varies with the spring index, which is the mean diameter divided by the wire diameter. The Wahl factor can be found with

$$K = \frac{4C - 1}{4C - 4} + \frac{0.615}{C} \tag{17-3}$$

where K = Wahl factor

C = spring index

A more convenient method is to use Fig. 17-4. This figure shows that K is always greater than 1, and as it is to allow for the increase in stress, Eq. (17-1) for corrected stress s_s' becomes

$$s_s' = \frac{8PDK}{\pi d^3} \tag{17-4}$$

The corrected stress is compared with the allowable stress to determine whether the spring is properly designed, but the stress s_s is used in the deflection equations.

If Eq. (17-1) is solved for P, and this is then substituted for P in Eq. (17-2), the following useful equation is produced:

$$F = \frac{D^2 N s_s \pi}{Gd} \tag{17-5}$$

Equations (17-2) and (17-5) can be modified to produce the deflection per coil:

$$f = \frac{D^2 s_s \pi}{Gd} \tag{17-6}$$

$$f = \frac{8PD^3}{Gd^4} \tag{17-7}$$

where f = deflection per coil.

Example 17-1 Find the shear stress in a spring of 3 in. outside diameter, made from 0.207 diameter wire, when subjected to a load of 100 lbs.

Solution $D = 3.000 - 0.207 = 2.793$. The stress is found using Eq. (17-1):

$$s_s = \frac{8PD}{\pi d^3} = \frac{8 \times 100 \times 2.793}{\pi 0.207^3} = 80,000 \text{ psi}$$

This is the uncorrected stress; the corrected stress is found by multiplying it by the correction factor K, which is determined by the use of Eq. (17-3) or Fig. 17-4. $C = D/d = 2.793/0.207 = 13.5$; thus, $K = 1.1$, and $s'_s = 80,000 \times 1.1 = 88,000$ psi.

Example 17-2 What is the total deflection of the spring in Example 17-1 if it has a total of 15 coils and has squared and ground ends, and the modulus of elasticity in shear is 11,500,000?

Solution For squared and ground ends, one coil is inactive at each end. Thus, $N = 13$. The total deflection is found using Eq. (17-2):

$$F = \frac{8PD^3N}{Gd^4} = \frac{8 \times 100 \times 2.793^3 \times 13}{11,500,000 \times 0.207^4} = 10.6 \text{ in.}$$

This deflection will produce the stress calculated in Example 17-1.

17-4 LENGTH

Each coil is limited in its deflection by the allowable stress; therefore, the length of a spring is dependent on the total deflection that is required. Several lengths are involved. The free length is the length when subjected to no load, the solid length is the length when compressed solid, the initial working length is the length when subjected to a specified load, and the minimum working length is the initial working length minus the working deflection. For a spring subjected to a static load, the minimum working length is the same as the initial working length.

The minimum working length must be greater than the solid length to prevent the coils from coming in contact with each other, since this would reduce the life of the spring. The minimum clearance between active coils at the minimum working length should not be less than 10 percent of the wire diameter. The solid length is $d(N + 3)$ for squared ends and $d(N + 2)$ for squared and ground ends. The free length may be a factor to consider, if provision is to be made in the mechanism to deflect the spring at assembly to the initial working length by adjusting the mechanism.

17-5 DESIGN STRESS

In determining the design stress for a particular spring, consideration must be given to the conditions under which the spring must operate. The effects of high or low temperature and corrosion will be discussed later.

If the operating conditions consist of a static load or a gradually applied load that is applied less than 10,000 times in the life of the spring, it is classified as light service. Severe service consists of considerable deflection at a rapid rate for long periods, such as a valve spring for an internal combustion engine. Average service is the same as severe service except that the total number of cycles is relatively small, such as the springs used in clutches or brakes.

Another factor that should be considered is the consequence of failure. If failure of the spring would endanger life or property, or if replacing it would involve considerable difficulty, it would be appropriate to regard it as subject to severe service even though the operating conditions indicate light service.

There are several ways of finding the design stress. One method is to divide the yield strength in shear by an appropriate factor. For light service this factor is 1.5, for average service it is 1.8, and for severe service it is 2.2. A check should be made to make certain that the stress does not exceed the yield strength in shear when the spring is compressed to the solid length. This check is made by substituting the difference between the free length and the solid length for F in Eq. (17-5) and solving for s_s.

17-6 MATERIAL

Though many materials are used for springs, round steel wire is used for by far the greatest number. Five of the most often specified steels will be discussed, along with two representative nonferrous metals. The first factor to be considered in selecting a steel is how the spring is to be made, which depends on the size of the wire to be used. An average spring with a wire diameter of less than 0.3625 will generally be cold-wound from hard-drawn or oil-tempered wire, but if the diameter is larger, the spring will be formed hot from hot-rolled bar.

Springs made from hard-drawn wire are not heat-treated after forming, because the physical properties of the steel are achieved by the method of producing the wire. The highest quality of hard-drawn wire is known as music wire (ASTM A228). It is used almost exclusively for springs with a wire diameter smaller than 0.024 and often for springs with a wire diameter to 0.130. The yield strength in shear for wire smaller than 0.020 is 180,000 psi; for diameters 0.020 to 0.051 it is 165,000 psi; and for wire larger than 0.055 it is 150,000 psi. The modulus of elasticity in shear is 12,000,000.

Hard-drawn spring wire (SAE 1350), which is not of the same quality as music wire, is widely used for low-stressed springs or for static applications, because its cost is less than one-third that of music wire. The yield strength in shear is 105,000 psi for 0.020 to 0.054 diameters, and 95,000 for larger wire. The modulus of elasticity in shear is 11,400,000.

Oil-tempered wire (ASTM A229) may be heat-treated to obtain its physical properties as part of the wire-making process, or it may be coiled while in the annealed condition and then heat-treated, which permits more severe forming than is possible with hardened wire. This wire is slightly more costly than hard-drawn spring wire. The yield strength in shear for 0.020 to 0.054 diameters is 140,000 psi, and for larger wire it is 120,000 psi. The modulus of elasticity in shear is 11,500,000.

The most common steel for hot-formed springs is SAE 1095. As the forming operation takes place at a temperature higher than the critical temperature of the steel, heat-treatment must be performed after forming. The yield strength in shear is 220,000 psi for diameters less than 1 in., and 100,000 psi for larger bar. The modulus of elasticity in shear is 10,500,000.

Stainless steel (type 302) is used where corrosive conditions prevail. It costs about 6 times as much as oil-tempered wire and is available in small diameters. The yield strength in shear is 125,000 psi for wire smaller than 0.020, 115,000 psi for 0.020 to 0.054 diameters, and 105,000 psi for larger wire. The modulus of elasticity in shear is 10,000,000.

Phosphor bronze is nonmagnetic and has good electrical conductivity. The yield strength in shear is 82,000 psi for wire 0.020 to 0.051 in diameter, and 74,000 psi for larger wire. The modulus of elasticity in shear is 6,000,000.

Inconel is used in applications involving high temperature or corrosion or both. It costs about 12 times as much as oil-tempered wire. The yield strength in shear is 108,000 psi for diameters 0.020 to 0.051, and 105,000 psi for larger diameters. The modulus of elasticity in shear is 10,500,000.

Standard wire sizes are shown in Table 17-1. Gage numbers are not given, because wire should always be specified in decimals. Music wire is available in sizes larger than that shown. Also, steel wire is available in sizes from $\frac{1}{32}$ to 1 in., in $\frac{1}{32}$-in. increments, and from 1 to 2 in., in $\frac{1}{16}$-in. increments.

17-7 SPRING RATE

The amount of deflection produced by a given load varies with each different spring. The term used to express the relation between the load and the deflection is spring rate; sometimes the term scale is used. The meaning is generally the load in pounds required to produce 1 in. of deflection; it is found by dividing the load by the deflection it produces. Theoretically, the spring rate is a constant, and if the deflection and load were plotted on a graph the curve would be a straight line, but the load-deflection curve of a real spring usually deviates from a straight line. This deviation is primarily the result of: a change in the number of active coils as the spring is compressed, eccentricity of the load, and friction at the ends.

TABLE 17-1   Standard Wire Sizes (Diameters in inches)

STEEL WIRE		MUSIC WIRE		NONFERROUS WIRE	
0.0230	0.1620	0.004	0.041	0.0285	0.1285
0.0258	0.1770	0.005	0.043	0.0320	0.1443
0.0286	0.1920	0.006	0.045	0.0359	0.1620
0.0317	0.2070	0.007	0.047	0.0403	0.1819
0.0348	0.2253	0.008	0.049	0.0453	0.2043
0.0410	0.2437	0.009	0.051	0.0508	0.2294
0.0475	0.2625	0.010	0.055	0.0571	0.2576
0.0540	0.2830	0.011	0.059	0.0641	0.2893
0.0625	0.3065	0.012	0.063	0.0720	0.3249
0.0720	0.3310	0.013	0.067	0.0808	0.3648
0.0800	0.3625	0.014	0.071	0.0907	0.4096
0.0915	0.3938	0.016	0.075	0.1019	0.4600
0.1055	0.4305	0.018	0.080	0.1144	0.5165
0.1205	0.4615	0.020	0.085		
0.1350	0.4900	0.022	0.090		
0.1483		0.024	0.095		
		0.026	0.100		
		0.029	0.106		
		0.031	0.112		
		0.033	0.118		
		0.035	0.124		
		0.037	0.130		
		0.039			

17-8   BUCKLING

A compression spring is a somewhat flexible column, and if it is rather slender it will buckle. The manner in which the spring is loaded also affects the tendency to buckle. If the load is applied by flat surfaces that rest against the spring, there is less of a tendency to buckle than if the ends rest against spherical surfaces. If the load is applied by flat surfaces and the free length is less than 5 times the mean diameter, buckling will not occur. If the free length is 6 times the mean diameter and the deflection is limited to 40 percent of the free length, or if the length is 8 times the mean diameter and the deflection is limited to 20 percent of the free length, buckling will not occur. If a spring is so long that buckling is likely, it can be placed over a rod or in a tube to prevent its buckling. However, consideration should be given to the facts that the spring rubbing against such a guide will cause wear and that the friction involved will affect its rate.

17-9   ENERGY ABSORBED BY A SPRING

The work done in compressing a spring, assuming that it is not under an initial load and that the deflection is proportional to the load, is the average load times the deflection, that is, $FP/2$. This amount of energy is not dissipated by a compression spring but is stored, and practically all of it will be released later.

17-10 SURGE

If the load is both rapidly and periodically applied, a spring may be subject to surging, which is a compression wave that travels along the spring. Surging can cause the mechanism associated with the spring to vibrate or flutter, and it may also cause the stress to be greatly increased and result in premature failure of the spring. A round steel wire compression spring has a natural frequency in cycles per second that can be approximated with

$$\phi = \frac{14,000d}{D^2 N} \tag{17-8}$$

If the natural frequency of the spring is less than 15 times the frequency of the load, consideration should be given to the possibility of surging.

17-11 EFFECT OF TEMPERATURE

Either high or low temperatures tend to affect springs adversely. At temperatures higher than 50°F below zero there is no effect of practical importance. At lower temperatures steel springs should not be subjected to impact loads, because of brittleness at such temperatures. Nonferrous spring materials are more satisfactory at lower temperatures. The slight increase in modulus of elasticity in shear at low temperatures is usually ignored. When a spring is compressed to a length that induces a reasonable working stress, either it will, as the temperature increases, support a smaller load at a specified length, or the deflection will increase if the load is held constant. These effects increase considerably with an increase in temperature. Further, for a given temperature that is relatively high, the effects will increase with time, but tend to level off after several days. The maximum working temperature for steel springs is about 300°F, providing the allowable stress is reduced to about 80 percent of the allowable stress at room temperature. For Inconel, the permissible temperature is over 700°F, with a similar reduction in stress. The maximum temperature for phosphor bronze is only one-half that of steel, and for stainless steel, the maximum is between that for steel and Inconel. The shear modulus decreases with an increase in temperature, and although the change is greater than with low temperature, for most applications the difference can be ignored.

17-12 SURFACE TREATMENT

The condition of the surface of a spring has an important effect on performance, especially if repeated loading is involved. A surface treatment that can greatly increase the fatigue life of a spring is shot peening. This process consists of propelling hard, small steel shot at the spring at high velocity, which causes microscopic dents in the metal. Thus, larger defects are smoothed out and a very thin layer of increased strength is produced due to the cold working. This process is not appropriate for wire diameters less than 0.0625 or for close-wound springs. Shot peening tends to reduce the corrosion resistance of the wire, but plating after shot peening is a good practice.

Corrosion must be considered in determining the material from which the spring is made or the protective treatment it is to be given. Even mild corrosive conditions can cause a great reduction in the life of a spring subject to repeated loading unless it is protected. Cadmium plating is often used to protect steel springs, but it must be done with care, and the springs should be properly baked immediately after plating to prevent hydrogen embrittlement. In applications involving a corrosive environment, nonferrous materials should be considered.

17-13 MANUFACTURE

As previously mentioned, springs are almost always made by a company that specializes in the manufacture of springs. It is a much better practice to specify exactly what is needed and let the company design the spring to satisfy the requirements than to have the company make it to a detailed specification. In providing the manufacturer with the requirements that the spring must meet, the following should be considered:

1. The number of springs required. Several different quantities may be specified, and a quotation will be made on each quantity.

2. If the spring is to work in a hole or over a rod, the diameter and tolerance of the rod or hole must be given.

3. If the solid length or free length must be limited, the maximum must be given.

4. The type of ends must be specified.

5. If the direction of coil is important, it must be specified.

6. If the ends must be square with the axis closer than 3°, the tolerance must be stated.

7. The environment that the spring is to work in should be described, such as temperature and whether subjected to corrosion, etc.

8. The spring rate, if important, is given, and the lengths between which this rate must be maintained are given. Also, a tolerance on the rate should be stated.

9. If the spring is used in a static application, the load that must be supported at a specified length must be given, with a tolerance on either the load or the length.

10. If the load varies, the frequency of application, the expected life, and either the two lengths or the two loads between which the spring will operate must be specified, with a tolerance on either the loads or the lengths; also, the initial working length and its accompanying load must be given.

It cannot be overemphasized that unless an item is important it should not be required. Also, that those items that are specified should be given as large a tolerance as possible. It would be appropriate to include as suggested or approximate: the wire diameter, the mean diameter, the free length, and the total number of coils, as an indication to the manufacturer what the designer has in mind, for many different springs could be made to satisfy the few items definitely specified.

17-14 EXTENSION SPRINGS

An extension spring is designed to resist a load that tends to make it longer. With little modification, the methods and data that were used in designing compression springs can be used in the design of extension springs. The two most important differences are initial tension and problems associated with the ends.

Extension springs are generally wound with the coils touching each other, and when initial tension is present, a considerable force must be applied before the coils are separated. Initial tension is produced by the manner in which the spring is wound. A spring may be made without initial tension; it may also be made with various amounts, but only when it is not subsequently heat-treated to develop the required strength.

The spring manufacturer prefers some initial tension in cold-wound springs; a minimum of 5 percent would be appropriate. The maximum amount of initial tension that can be easily accomplished is about 20 percent. A spring cannot carry more load nor does it have a lower stress because of initial tension. The importance of initial tension to the designer is that the load referred to as initial tension must be applied before deflection begins. Thus, the total load minus the initial tension is used in calculating the deflection with Eq. (17-2), and the total load is used in calculating the stress with Eq. (17-1).

An extension spring must have some means of attaching it to the mechanism that it is a part of. When this is accomplished by forming loops or hooks in the ends of the wire of which the spring is made, they become the part of the spring most vulnerable to failure. This is because of the large stress induced at the small radii that form the end, especially at the transition of the last coil to the loop. This higher stress can be allowed for by using a larger factor in determining the allowable stress than was used

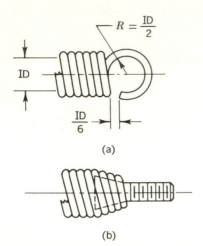

(a)

(b)   **Fig. 17-5 Extension spring ends.**

for compression springs. For light service this factor is 1.8, for average service 2.3, and for severe service 2.8.

A commonly used type of end is shown in Fig. 17-5a. It should be kept in mind that ends differing from that shown will considerably increase the cost, as they cannot be formed automatically in the machine that coils the spring.

The problem of ends can be solved in another way, and at the same time other features can be provided that may be desirable, such as adjustment or the freedom of a swivel. An example of such an end is shown in Fig. 17-5b. When the ends are provided for in such a manner, the factor for determining the allowable stress can be slightly reduced.

In determining the deflection, only the active coils are used in the equation. For ends that consist of separate parts, the coils in contact with these ends are not active. For ends formed as part of the spring, the number of active coils may have to be increased to allow for the deflection that takes place in the loop. If the loop is small, the deflection is negligible, but if the loop is as shown in Fig. 17-5a, the deflection due to each loop is equal to one-half the deflection of a coil. Thus, if the spring has such a loop at each end, the number of active coils is the number in the body of the spring plus 1.

To ensure that the spring will fit properly with the mechanism associated with it, the manufacturer must be given a drawing showing the ends that are required. A more appropriate procedure would be to supply the manufacturer with a drawing showing the provision that has been made for attaching the spring. As to the additional information that should be given to the manufacturer, items 1, 5, 7, 8, 9, and 10 of Sec. 17-13 are applicable, plus the following:

1. If there is a limit to the size of the outside diameter, it must be specified.

2. The maximum length without set that is required should be given. This may be dictated by assembly of the mechanism.

3. The initial tension with a tolerance that should be at least 10 percent.

As discussed in Sec. 17-13, an item should not be specified unless it is important. Again, it would be appropriate to include the suggested wire diameter, mean diameter, the number of active coils in the body, and the length when supporting no load.

17-15 CONCENTRIC SPRINGS

Concentric helical springs are not a different type of spring but a different arrangement. Usually two regular compression springs are placed on the same axis, one inside the other, to produce this arrangement. The capacity to handle a large load in a small space, and the added safety of two springs, are two of the advantages of this arrangement. However, the most common reason for its use is that two springs often eliminate the undesirable surge that would occur if only one spring were used. Each spring should have about the same maximum stress. This can be accomplished by making the spring indexes equal. Adjacent springs should be wound in opposite directions to avoid the possibility of the coils locking.

17-16 FLAT SPRINGS

Springs made from sheet metal are called flat springs, even though they are often formed into rather intricate configurations. Because there is no standard configuration, there is no generalized method of design that may be used. Equations for the determination of stress and deflection of various types of beams can often be used in the design of such springs. Sometimes they are developed in the shop by trial and error. Factors that should be given careful consideration, whichever method is used to produce the design, are:

1. There should be reasonably large fillet radii in the blank before forming.

2. Holes should not be too close to the edge or to each other.

3. There should be relatively large bend radii in forming.

4. Close angular relationships to be accomplished by bending should not be required.

Another type of spring that is similar to those just discussed, and to which most of the discussion applies, is one that differs only in that it is made of wire.

17-17 LEAF SPRINGS

A leaf spring is an assembly of several narrow strips placed one on top of another and securely clamped together. Millions of automobiles have been made with such springs.

The greatest advantage of this type of spring is that it can be used as a structural member in addition to its function as a spring. An example of this is an automobile that uses these springs at the rear axle. Two such springs serve as the only means of maintaining position and alignment of the rear wheels.

Largely because this type of spring can be made to serve as a structural member, its design becomes more complex than at first appears, especially if it is to be used in an application where fatigue is a factor.

17-18 TORSION SPRINGS

Torsion springs are generally helical springs arranged as shown in Fig. 17-6 to offer resistance to a moment that is similar to that used in coiling the spring; that is, the load tends to wind up the spring. A torsion spring may be used in an application where the tendency is to unwind the spring, but this is undesirable, especially if it is cold-wound. In designing for a torsion spring, provision should be made for a rod to fit inside the coils and serve as a guide to prevent buckling. The ends should be as simple as possible both to facilitate manufacture and to reduce the stress concentrations that accompany intricate ends. When a guide is used, the friction between the spring and the guide results in a less consistent performance than can be obtained with other spring forms, but this is not important in most applications.

Fig. 17-6 A torsion spring.

17-19 TORSION BAR SPRINGS

A torsion bar spring is a straight bar, so arranged that the applied load causes the bar to be twisted. The bar is usually of solid circular cross section, though tubular or square, even rectangular, cross sections are sometimes used. The usual equations for stress and deflection for members in torsion may be used in design. This form of spring is used in a very wide range of sizes, also in applications where consistency of performance is essential, as in instruments. In the larger sizes, the most efficient means of applying the load is by means of splines. The diameter at the ends should be increased, and there should be a gradual transition from this larger diameter to the main body of the bar.

17-20 CONICAL SPRINGS

A conical spring is one that has the coil diameter at one end considerably smaller than at the other end, as shown in Fig. 17-7. This type of spring has several characteristics that may be desirable in some instances. This spring can be arranged to compress to a much smaller solid length than the ordinary compression spring. Also, as the larger coils bottom or become inactive, the spring becomes stiffer, that is, the spring rate increases. The use of this type is limited, because it is much more difficult to produce a satisfactory spring of this type than the usual compression spring; also, as the coils become inactive, the stress in the remaining active coils greatly increases.

17-21 RING SPRINGS

A ring spring is an assembly of rings, each with an appropriate cross section, as shown in Fig. 17-8. When a load is applied, a tensile stress is

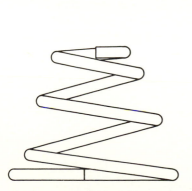

Fig. 17-7 A conical spring.

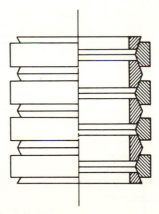

Fig. 17-8 A ring spring.

induced in the outer rings and a compression stress in the inner rings. There will be considerable friction as the rings slide on each other, which also resists the applied load. When the load is reduced, the energy stored in the stretched and compressed rings is released. But because of the friction of the rings sliding on each other as the spring extends, a great deal less energy is released than if a common compression spring was used. This damping action is the most desirable characteristic of this type of spring. Other characteristics are: the configuration of the cross section of the rings is very important; the deflection is relatively small; the loads are usually measured in tons; and there is considerable wear due to the rings rubbing on each other.

17-22   DISK SPRINGS

Disk springs (also called Belleville springs or Belleville washers) are an assembly of cone-shaped circular disks with a hole in the center, as shown in Fig. 17-9. When the load is applied, the disks tend to flatten out. The characteristics of this spring are a high load-carrying capacity for the space required, a relatively small deflection, and load-deflection characteristics that can be varied by the manner in which the disks are stacked. If the disks are all stacked facing the same way (rather than alternating, as in Fig. 17-9) considerable friction and thus damping results. A guide in the form of a rod at the center or a tube in which the disks are placed should be provided. An appropriate clearance must be allowed between the disks and either the rod or the tube.

17-23   FLAT SPIRAL SPRINGS

Flat spiral or motor springs are the most efficient form of spring for storing energy. This is the type that is used in clocks and phonographs. This spring is in the form of a relatively thin steel ribbon or strip wound in a spiral. One end is connected to an arbor, the other to the case that restrains it when unwound. Though this spring is designed to deliver a torque, it may, through the use of a suitable mechanism, be used to exert a push or

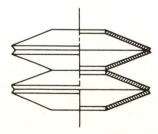

Fig. 17-9   Disk springs.

pull, often resulting in a more compact assembly than a compression or extension spring would allow.

Another form of spiral spring is the type where the turns do not come in contact with each other as they do in the motor type. The advantage of this type is that, as the turns do not come in contact, there is no unpredictable friction. This results in consistency of performance that recommends their use in instruments of the rotating pointer type.

QUESTIONS

17-1 What are some of the basic purposes for which springs are used?

17-2 Describe the various types of ends produced on compression springs.

17-3 What are the advantages and disadvantages of an end that is squared and ground?

17-4 What kind of stress does the load produce in a compression spring?

17-5 What determines the length of a compression spring?

17-6 Explain free length and solid length.

17-7 Why are springs that are made from hard-drawn wire not heat-treated?

17-8 Define spring rate.

17-9 Discuss what can be done about the tendency of a long spring to buckle.

17-10 Why is surging of a spring usually undesirable?

17-11 What is the advantage of shot peening a spring?

17-12 What process should follow the cadmium plating of springs?

17-13 Explain initial tension.

17-14 Discuss briefly the problem of ends for extension springs.

17-15 What is the most common reason for the use of concentric springs?

17-16 What is the greatest advantage of a leaf spring?

17-17 What is the difference between a torsion spring and a torsion bar spring?

17-18 Describe a conical spring.

17-19 Describe a ring spring.

17-20 What is the most desirable characteristic of a ring spring?

17-21 Describe a disk spring.

17-22 For what situations is a disk spring most appropriate?

17-23 Describe two types of spiral springs.

PROBLEMS

Note: After solving some of these problems, study of Design Example 6 in Chap. 19 and completion of one of the following is *recommended:* Project 14 if Project 13 has been completed, Project 15 if Project 13 has not been completed.

17-1 What shear stress is produced in a compression spring of 2 in. outside diameter, made from 0.162 wire, when subjected to a load of 150 lb?*

17-2 What is the corrected stress for Prob. 17-1?

17-3 If oil-tempered wire is used for the spring in Prob. 17-1, what is the deflection per coil produced by a load of 10 lb?*

17-4 If the spring in Prob. 17-1 has squared and ground ends and 12 active coils, what is the solid length?

17-5 What is the design stress for 0.026 music wire used for a compression spring intended for average service?*

17-6 If the spring in Prob. 17-1 has 8 active coils, what is the natural frequency of the spring?

17-7 If the spring in Prob. 17-1 has a total of 13 coils and squared and ground ends, what is the total deflection when the stress is 90,000 psi?

17-8 What is the energy absorbed by the spring in Prob. 17-7?

17-9 An extension spring made of 0.207 hard-drawn wire has an outside diameter of 2.75, an initial tension of 15 percent, and 18 active coils. What are the corrected stress and the total deflection, when it is subjected to a load of 200 lb?*

17-10 What is the design stress for 0.283 oil-tempered wire, to be used in an extension spring subjected to severe service?

SELECTED REFERENCE

Wahl, A. M.: "Mechanical Springs," McGraw-Hill Book Company, New York, 1963.

MISCELLANEOUS MACHINE ELEMENTS

18-1 POWER SCREWS

A power screw is a screw and nut arrangement used for transmitting power or motion, and in some instances for converting one form of motion to another. In the most common arrangement, the screw rotates and the nut moves along the screw. The nut may rotate, causing the screw to move along its axis, or the nut may be fixed and the screw may both rotate and move axially. Other arrangements are also possible.

The square-thread form, shown in Fig. 9-2, is theoretically the most efficient for use in power screws; however, it costs more to manufacture and it is difficult to provide a means to compensate for wear. The most frequently used thread is the Acme thread, shown in Fig. 9-2. It is economically produced, it may be used with a split nut to bring about engagement with a rotating shaft, and incorporating a means to compensate for wear and to eliminate backlash is easy. Also, because of the superior surface that can be produced, it is practically as efficient as the square thread.

Regardless of the thread form, there is considerable friction between the screw and the nut. There is also friction at the thrust collar. This can be greatly reduced by using a ball or roller thrust bearing. Efficiency increases with an increase in the lead; thus, multiple threads are often used. However, the mechanical advantage decreases with an increase in lead.

As the lead is increased, a point is reached where the screw is not self-locking. If a screw that was not self-locking was used in a jack, removing the torque that raised the load would permit the load to descend.

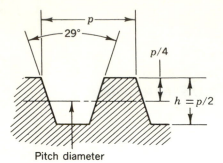

Pitch diameter   **Fig. 18-1 Acme thread dimensions.**

Self-locking may or may not be an advantage; it depends on the individual situation.

Acme threads have not been standardized to the extent that screw fasteners have been; however, the dimensions and proportions for some selected sizes that are recommended for the usual applications are given in Fig. 18-1 and Table 18-1.

TABLE 18-1   Dimensions of Acme Screw Threads

DIAMETER, IN.	THREADS PER IN.
0.25	16
0.50	10
0.75	6
1.00	5
1.25	5
1.50	4
1.75	4
2.00	4
2.50	3
3.00	2
4.00	2
5.00	2

The torque required to exert a given force with an Acme screw is

$$T = \frac{Qd}{2}\left(\frac{\cos \alpha \tan \lambda + \mu}{\cos \alpha - \mu \tan \lambda}\right) \qquad (18\text{-}1)$$

where T = torque, lb-in.
Q = load or force, lb
d = pitch diameter, in.
λ = lead angle
μ = coefficient of friction

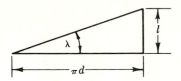

Fig. 18-2 Lead angle.

The relationship between the lead l and the pitch diameter d is shown in Fig. 18-2. The coefficient of friction for high-quality work that is properly lubricated is about 0.15 for starting and 0.1 for running. For poor-quality work and poor lubrication, it is about 0.2 for starting and 0.15 for running.

The torque required to overcome the friction at the thrust collar is found with

$$T_c = \frac{\mu Q d_c}{2} \tag{18-2}$$

where T_c = torque, lb-in.

d_c = mean diameter of thrust collar, in.

The coefficient of friction for a steel collar on a bronze thrust bearing is about 0.1 starting and 0.08 running. If a ball or roller thrust bearing is used, the friction is so small relative to that of the screw that it may be ignored.

The total torque required to exert the force Q is the sum of the torque from Eq. (18-1) and the torque from Eq. (18-2).

Example 18-1 What torque is required to exert a 2000 lb force with a high-quality 1.0-in.-diameter Acme thread? The mean diameter of the bronze thrust collar is 1.38. The running torque, not starting torque, is required.

Solution Referring to Table 18-1 and Fig. 18-1, the pitch = 1.00/5 = 0.20 in. The distance from the outside diameter to the pitch diameter is 0.20/4 = 0.05 in. Thus, the pitch diameter is 1.00 − 0.10 = 0.90 in.

$$\tan \lambda = \frac{0.20}{\pi 0.90} = 0.071$$

$$\cos \alpha = \cos 14.5° = 0.968$$

The running coefficient of friction is 0.1 for the screw and 0.08 for the collar. The torque required for the screw is found with Eq. (18-1):

$$T = \frac{Qd}{2}\left(\frac{\cos \alpha \tan \lambda + \mu}{\cos \alpha - \mu \tan \lambda}\right)$$

$$= \frac{2000 \times 0.9}{2}\left(\frac{0.968 \times 0.071 + 0.1}{0.968 - 0.1 \times 0.071}\right) = 158 \text{ lb-in.}$$

Fig. 18-3   A ball bearing screw. (*Saginaw Steering Gear Div., General Motors Corp.*)

The torque required to overcome friction at the collar is found with Eq. (18-2):

$$T_c = \frac{\mu Q d_c}{2} = \frac{0.08 \times 2000 \times 1.38}{2} = 110 \text{ lb-in.}$$

Thus, the total torque required is $158 + 110 = 268$ lb-in.

A unique power screw assembly is the ball-bearing screw developed by the Saginaw Steering Gear Division of the General Motors Corporation. Such an assembly is shown in Fig. 18-3. The balls roll between the nut and

Fig. 18-4   A large stripping shovel. (*Bucyrus-Erie Co.*)

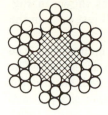

Fig. 18-5   A 6 × 7 wire rope cross section.

the screw in the same manner as a ball bearing. At a lead angle of 2° the efficiency of such a unit is over 90 percent, whereas the efficiency of an Acme screw is about 25 percent.

18-2   WIRE ROPE

Wire rope has great strength and long life when used in properly designed applications. It may also be wound on a drum, which permits the design of relatively simple hoists, cranes, etc. An excellent example of the use of wire rope in a machine is shown in Fig. 18-4. This shovel is one of the world's largest. It has a dipper capacity of 140 cubic yards and is powered by 52 electric motors from ¼ to 3000 hp.

Wire rope is made by twisting a number of wires into a strand and wrapping the strands about a core. The core may be hemp or wire. A wire core has somewhat greater strength and is much more resistant to crushing. The construction is indicated by two numbers; the first specifies the number of strands, the second the number of wires in each strand. Thus,

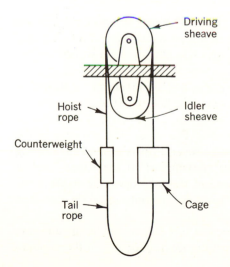

Fig. 18-6   An elevator system.

a 6 × 7 rope has 6 strands with 7 wires in each strand, as shown in Fig. 18-5. The size of a rope is the diameter of a circle that will just contain the rope.

Flexibility is achieved by using a large number of small wires. The 6 × 7 rope is stiff and is used mainly for fixed wires, such as rigging. The 6 × 19 construction is often used where flexibility is required, such as for hoists. The 6 × 37 is extra flexible and is used where rather sharp bends are necessary, but it does not have the resistance to abrasion of the stiffer ropes.

The strength of a rope is always less than the theoretical sum of the strength of the wires of which it is made. The modulus of elasticity is from one-half to one-third that of the metal of which the wire is made, because of the sliding of the wires on each other as the load is applied.

When a wire rope passes around a pulley (called a sheave when dealing with wire rope) or a drum, the wires are bent, a wire is forced against those adjacent to it, and there is motion between the wires as they change from the curved condition at the bend to the straight condition. The wires also rub against the sheaves. These actions will eventually cause failure even in a properly designed system. The smaller the sheave relative to the diameter of the rope, the greater will be these undesirable effects.

The size and shape of the groove in the sheave is important. If it has a flat bottom or too large a radius, the rope will tend to lose its circular cross section, which increases the fatigue effects. If the groove has too small a radius, the rope will tend to become wedged in the groove, which will cause unnecessary wear of both the rope and the sheave. The proper radius for the groove is: the radius of the rope plus 0.015 for ropes smaller than ⅜ in., the radius plus 0.03 for ropes from ⅜ to 1¼, and the radius plus 0.06 for larger ropes.

When a wire rope is used with a drum, it is desirable to have grooves similar to those described for a sheave, though often this is not feasible, and the rope is wound on top of the rope that has already been wound on the drum.

In some instances, a wire rope system can be arranged as shown in Fig. 18-6, in which case a drum is not used. The action of the driving sheave is much the same as that of the drive pulley in a belt drive. The diameter should be at least 65 times the diameter of the rope. The purpose of the idler sheave is to permit the rope to make several turns around the driving sheave, which will increase the power transmitted to the rope. Such a sheave is not always used. The purpose of the tail rope is to balance the weight of the hoist rope in order to keep the magnitude of the counterbalancing force more nearly constant.

The appropriate design factor, which in this case would be the ultimate strength of the rope divided by the working load, depends on the application. For mine hoists, 3 could be acceptable; whereas, for a passenger elevator, 12 may be required by law. For derricks and cranes, about 5 is appropriate. In some applications several individual ropes may be best. For passenger elevators, a minimum of 4 is often required by law. The total load is distributed among the ropes.

The stress induced in the outer wires by bending is found with

$$s_b = \frac{E_r d_w}{D} \tag{18-3}$$

where s_b = bending stress, psi

E_r = modulus of elasticity of the rope

d_w = wire diameter, in.

D = diameter of the sheave, in.

It would be more convenient to allow for bending by finding an equivalent load that would produce the same stress, for this load could be added to the other loads to be resisted by the rope. This may be found by multiplying the stress by the cross-sectional area of the metal in the rope. Thus, Eq. (18-3) becomes

$$F_b = \frac{A E_r d_w}{D} \tag{18-4}$$

where F_b = equivalent load, lb

A = cross-sectional area, sq in.

The force involved in accelerating both the load and the rope must be added to the load to be resisted by the rope. When lifting vertically, this is

$$F_a = \frac{aW}{32.2} = \frac{2WL}{32.2t^2} \tag{18-5}$$

where F_a = force causing acceleration, lb

W = weight, lb

L = distance traveled during acceleration, ft

a = acceleration due to force F_a, fps^2

t = time during which acceleration takes place, sec

32.2 = acceleration due to gravity

Wire rope is manufactured in a number of forms and from various kinds of steel. Therefore, the strength, minimum diameter of sheave, and so on, will vary with the details of construction. The approximate properties and limitations of only one representative type, the 6 × 19 type, can be given here. The ultimate strength in pounds is 72,000d^2, where d is the diameter of the rope in inches. The weight per foot in pounds is 1.6d^2. Sizes are as follows: ¼ to ⅝ in ¹⁄₁₆ increments, ⅝ to 2¼ in ⅛ increments, 2½, and 2¾. The recommended minimum sheave diameter is 30d, but a diameter of $45d$ is more desirable. The wire diameter $d_w = 0.063d$; and the cross-sectional area of the metal $A = 0.38d^2$. The modulus of elasticity is 12,000,000.

18-3 FLYWHEELS

Some power sources, such as internal-combustion engines, produce energy during a small portion of the cycle. A flywheel is used to smooth out these fluctuations and to make the flow of energy uniform. Some machines, such as punch presses, require energy only for a small portion of the cycle, and a flywheel is used as a reservoir for storing energy to provide for these peak periods. This permits a much smaller power source than would otherwise be required.

The kinetic energy in a flywheel is

$$KE = \frac{Wv^2}{2g}$$

where KE = kinetic energy, ft-lb
$\quad\quad W$ = weight of the flywheel, lb
$\quad\quad\; v$ = velocity of the center of mass, fps
$\quad\quad\; g$ = acceleration due to gravity

The change in kinetic energy is

$$\Delta KE = KE_1 - KE_2$$

or $\quad \Delta KE = \dfrac{Wv_1{}^2}{2g} - \dfrac{Wv_2{}^2}{2g}$

or $\quad \Delta KE = \dfrac{W}{2g}(v_1{}^2 - v_2{}^2)$ $\hfill$ (18-6)

where the subscripts 1 and 2 represent the initial and final velocities, respectively.

The average velocity is

$$v = \frac{v_1 + v_2}{2} \tag{18-7}$$

where v = average velocity, fps.

The acceptable variation in angular velocity depends on the application. The coefficient of regulation is used to specify the magnitude of the variation in angular velocity. It is found with

$$C_f = \frac{v_1 - v_2}{v} \tag{18-8}$$

where C_f = coefficient of regulation. Equation (18-6) may be written

$$\Delta KE = \frac{W(v_1 - v_2)(v_1 + v_2)}{2g} \tag{18-9}$$

Equation (18-8) may be written

$$C_f v = v_1 - v_2$$

Equation (18-7) may be written

$$2v = v_1 + v_2$$

Substituting $C_f v$ for $v_1 - v_2$ and $2v$ for $v_1 + v_2$ in Eq. (18-9),

$$\Delta KE = \frac{W(C_f v)(2v)}{2g}$$

or $\Delta KE = \dfrac{W C_f v^2}{g}$ (18-10)

Since practically all the effect of a flywheel that has a rim and spokes is contributed by the rim, the effect of the hub and spokes is often ignored. If the rim thickness is small relative to the diameter (as is usually the case), the center of the rim can be taken as the center of mass. The acceleration due to gravity is usually taken as 32.2. Thus, Eq. (18-10) becomes

$$W = \frac{32.2 \, \Delta KE}{C_f v^2} \tag{18-11}$$

where W = weight of rim, lb

v = average velocity of the center of the rim, fps

Several typical values for the coefficient of regulation are: crushers 0.2, geared drives 0.02, dc generators 0.002, reciprocating pumps and compressors 0.04, and punch presses 0.08. A flywheel may also serve as a pulley, in which case the width, and even the diameter, may be dictated by the belt to be used.

Example 18-2 Determine the size of a cast-iron flywheel for a small rock crusher. The diameter is about 2 ft. It turns 200 rpm. The energy to be supplied is 500 ft-lb. The rim must be at least 4 in. wide to accommodate belts.

Solution The velocity, as found with Eq. (14-4), is $0.262 \times 24 \times 200 = 1260$ fpm, but since the velocity must be in fps, divide by 60. Thus, the velocity is 21 fps. The weight of the rim is found with Eq. (18-11):

$$W = \frac{32.2 \, \Delta KE}{C_f v^2} = \frac{32.2 \times 500}{0.2 \times 21^2} = 182 \text{ lb}$$

Cast iron weighs 0.27 lb/cu in. Thus, the rim must contain 675 cu in. Say that the rim was 26 in. outside diameter and 22 in. inside diameter. The area would be 0.785 $(26^2 - 22^2) = 149$ sq in. Therefore, the width of the rim would be $675/149 = 4.53$ in., which is a little more than the required minimum and is thus satisfactory.

18-4 VIBRATION DAMPERS

In a discussion of vibration two frequencies are important. One is the frequency of the disturbing force, which is related to the operating speed of the machine and often cannot be changed by the designer. The other is the natural frequency of that which is caused to vibrate by the disturbing force. This natural frequency is independent of the operating speed, but it is largely dependent on the mass of the vibrating item and the stiffness of the members that support it.

Vibration is especially objectionable when the frequency of the disturbing force coincides with the natural frequency. This is known as resonance. When resonance occurs, the amplitude of the vibration increases and a force larger than the original disturbing force develops.

The best way to avoid the undesirable effect of vibration is to eliminate the disturbing force. This is often impossible. Another way is to balance the disturbing force. When this is not feasible, the natural frequency should be changed sufficiently to prevent resonance. This can sometimes be accomplished by increasing or decreasing the stiffness of the supporting members. It can also be achieved by use of a vibration mounting.

Whenever vibration is present and undesirable, a vibration mounting should be employed. Such a mounting should be flexible and provide a damping action. A spring is often used to achieve the required flexibility, but it provides practically no damping. A friction material rubbing against a surface can be arranged, as shown in Fig. 18-7, to provide the damping.

Rubber is a very effective material for use in a vibration mounting, because of its flexibility and damping characteristics. Rubber is practically

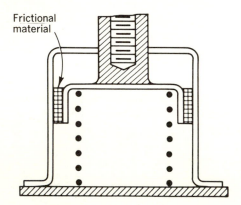

Frictional material

Fig. 18-7 A vibration damper.

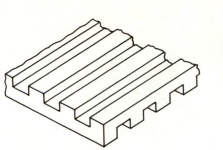

Fig. 18-8 A vibration isolating pad.

Fig. 18-9 A vibration mount.

incompressible, and when subjected to a compressive load it will expand in a direction perpendicular to the load. Thus, for a rubber pad to be effective as a vibration mounting, this expansion must not be restricted as it would be for a solid sheet. A form such as that shown in Fig. 18-8 is appropriate. The pressure on the pad should be limited to about 50 psi. Rubber used in this manner is effective in isolating high-frequency vibrations that create noise, but not low-frequency vibrations.

When rubber is subjected to a shear load, there is no expansion, and the flexibility is greater. Thus, most vibration mounts are arranged to place the rubber in shear. Such a mounting is shown in Fig. 18-9. The rubber is bonded to the two tubes. Such a mounting is placed at each point at which the vibrating unit is supported. Such a mounting may also be used to mount a sensitive unit and thus protect it from vibration. The working stress of the rubber in shear should not exceed 50 psi.

One of the reasons for using a flexible coupling is to reduce the transmission of torsional vibration. Couplings that incorporate a resilient material are more effective for this purpose.

When parts are designed and manufactured so as to be in balance, it must be remembered that unbalance of the assembly may result from the dimensional tolerances, such as a loose fit between a flywheel and a shaft.

18-5 SEALS AND GASKETS

A number of special situations will not be included in this discussion, such as high-speed steam turbines, high-pressure hydraulic systems, and engine pistons. Seals may be classified on the basis of whether they are static or dynamic. They may also be classified as exclusion seals, when they serve to exclude dirt, or as retention seals, when they prevent the leakage of oil or grease. A seal may be required to perform its function with either reciprocating motion or rotation.

Fig. 18-10 A single-lip spring-loaded synthetic rubber oil seal. (*Victor Divisions, Dana Corp.*)

Factors that should be considered in providing a seal include:

1.   The rubbing velocity — When the shaft rotates this is dependent on both the diameter and rpm of the shaft.

2.   The nature of that which is to be excluded or retained — Dirt, liquid, grease, oil.

3.   The pressure to which the seal is subjected — Both the magnitude and whether it is internal or external.

4.   The temperature to which the seal is subjected — The temperature may be increased by the rubbing of the seal.

5.   The hardness and roughness of the surface against which the seal rubs.

Except in instances where sealing is relatively unimportant, and where the seal is subjected to very light service, it should not be required to serve both retention and exclusion functions. A simple seal that is appropriate for light service is a felt ring that fits in the housing and rubs on the shaft. If a ring cannot be used, a felt strip fitted in a groove can be substituted.

A lip seal has a lip that rubs against the shaft, as shown in Fig. 18-10. It will be noted that the area in contact with the shaft is very small. These seals may be made from leather or from any one of a number of elastomers. The elastomers generally allow less leakage, but the leather is recommended where the seal may run dry for short periods. These seals are commonly referred to as oil seals.

Exclusion seals that are similar to lip seals are commonly called wipers.

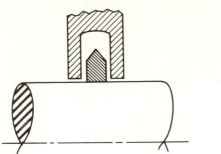

Fig. 18-11   A slinger seal.

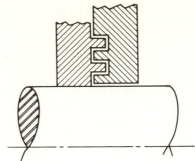

Fig. 18-12   A labyrinth seal.

Another exclusion device is referred to as a scraper. It has a metal lip and is used to remove material such as mud from a reciprocating rod. A wiper is usually used behind a scraper to exclude fine particles and liquid.

The friction of a lip seal or wiper used with a high-speed shaft may generate enough heat to cause deterioration of the seal. In such a situation, the device shown in Fig. 18-11, known as a slinger, may sometimes be used. There is no rubbing, and centrifugal force tends to throw the material outward, thus preventing it from passing through the seal. It may be used either as an exclusion seal or as a retention seal. Another arrangement that does not involve rubbing, and that can be used either for exclusion or retention, is the labyrinth seal. An example of this is shown in Fig. 18-12. The circuitous path greatly interferes with the passage of any material. A labyrinth seal and a slinger are sometimes combined.

When a static seal is required between flat parallel surfaces, a gasket

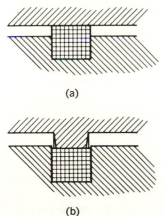

(a)

(b)

Fig. 18-13   Gasket installations.

is often used. For a gasket to maintain a tight joint, the surfaces must be flat and rigid, and the bolts must be tightened enough to cause local yielding of the gasket at the high spots on the surfaces. If a gasket is to fit in a groove, it should not protrude as shown in Fig. 18-13a, but it should be used with a tongue as shown in Fig. 18-13b.

If the joint is subjected to high temperature, the effect of thermal expansion and relaxation of the bolts must be considered. If a high internal pressure is involved, the bolts should be preloaded.

18-6 CAMS

A cam is a convenient means of transforming one form of motion to another, usually rotary motion to reciprocating motion. Because of the great number of shapes that are possible, many different motions can be produced. The determination of the shape or profile of a cam is covered in texts on kinematics. Several points that may not be treated in such a text are as follows: It may be impossible to keep the follower against the cam, because of the inertia of that which is connected to the follower or because of the rapid movement that is expected. Deformation of the parts between the cam and the point where the motion is required may cause the motion to be considerably different than the profile of the cam would indicate. The force exerted by a cam acts perpendicular to its surface, which is often not the direction in which the follower is to move. If the follower is sliding in a guide, this will cause the follower to be forced against the guide. The friction and wear that this causes is undesirable; thus, the angle between the force and the direction of motion of the follower should be as small as possible. A flat surface or cylindrical roller on the follower will produce line contact with the cam. A combination of cam width and hardness must be provided to withstand the load involved. A spherical end on the follower should be avoided, because it would have point contact with the cam.

18-7 FRAMES

A frame is the member that serves to maintain the proper relative location of the various components. It may be a simple item made by welding several pieces of steel together, or a complex casting that is extensively machined. The frame is too often accepted as possessing the qualities of an ideal frame, and such things as deflection under load and dimensional changes with variation in temperature are overlooked.

It is usually impossible to calculate the stress in a frame accurately, because of the many loads to which it is subjected and the fact that the configuration is largely determined by the various nonstructural functions that it is to perform. Thus, the size and configuration of the features are dependent on experience and judgment, keeping in mind that many of the loads act simultaneously and that deflection rather than strength is usually the determining factor.

18-8 STANDARD UNITS

The desirability of using standard items has been discussed, and many such items are familiar. Thus, in the design of a unit to reduce the speed of a motor, standard bearings, gears, belts, pulleys, chains, and sprockets would be considered. However, a standard unit known as a gear reducer could be purchased. A motor and gear reducer in a single unit is available in many sizes. Such a unit is shown in Fig. 18-14. A similar unit that permits varying the speed while the motor is running is also available. An example is shown in Fig. 18-15.

A beginning designer should become aware of the existence of such standard units in order to avoid designing something that it would be

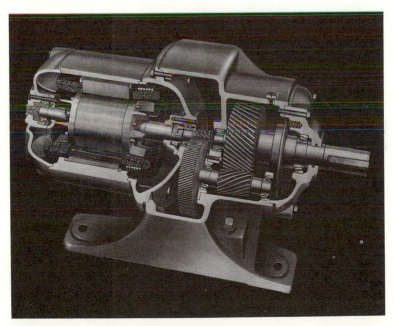

Fig. 18-14 A standard unit incorporating both motor and gear reducer.
(*U. S. Electrical Motors, Div. of Emerson Electric Co.*)

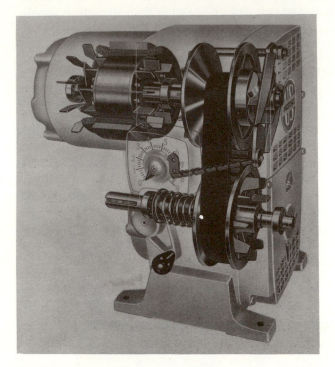

Fig. 18-15 A variable-speed drive and motor combination.
(*U. S. Electrical Motors, Div. of Emerson Electric Co.*)

better to purchase. One of the best ways of gaining this awareness is to read the advertisements in the trade magazines.

When a standard unit is to be used, it is the responsibility of the designer to become thoroughly familiar with both the unit and the manner in which it is to be used, in order to make certain that it is suitable, that its capacity is adequate, and that it does not have costly features and capabilities which are not required by the application.

18-9 OTHER MECHANISMS

There are many mechanisms with which a designer should be familiar. A ratchet wheel and pawl, as shown in Fig. 18-16, can be used to prevent a shaft from rotating in the reverse direction. These same elements can be combined, as shown in Fig. 18-17, to accomplish turning the shaft by an oscillating motion of the lever.

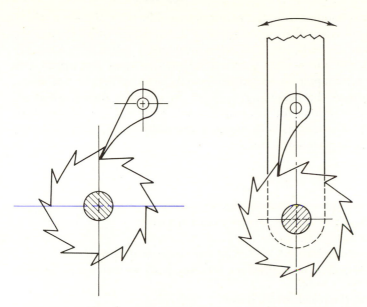

Fig. 18-16   A ratchet wheel and pawl.   **Fig. 18-17   A ratchet drive.**

The mechanism shown in Fig. 18-18, known as a Geneva drive, causes the shaft to which the slotted disk is attached to make 1 revolution for each 5 revolutions of the other shaft. A different number of slots will produce other ratios.

A differential screw is an arrangement that provides very small axial motion per revolution using threads with a reasonably large pitch.

18-10   CONCLUSION

During the discussion of the principles and concepts involved in mechanical design and the more common machine elements, it has no doubt become obvious that the more machine elements, mechanisms, and units

Fig. 18-18   A Geneva drive.

that a designer is familiar with, the better prepared he is for the work which he is to do. It is essential for a designer to develop the ability to find the information that he needs, for the field of mechanical design is so vast that he cannot expect to know the answers to all the questions that arise. The discussions in this chapter are incomplete, partly because there is not space, but mainly to start the beginning designer on a search for information. This process of searching for information and becoming acquainted with it will continue for as long as he is engaged in making a significant contribution in the field of mechanical design.

QUESTIONS

18-1 Compare the square and Acme threads for use as power screws.

18-2 Discuss the effect of lead on the performance of a screw.

18-3 Describe a ball-bearing screw.

18-4 A wire rope is designated 6 × 19. What is the meaning of the numbers?

18-5 Why is the modulus of elasticity of a wire rope much less than that of the metal from which it is made?

18-6 What is the purpose of an idler sheave?

18-7 Describe the two major uses of a flywheel.

18-8 What is a coefficient of regulation?

18-9 What effect may a belt have on the design of a flywheel?

18-10 In a discussion of vibration, what are the two most important frequencies?

18-11 What is resonance?

18-12 Discuss rubber as a material for use in a vibration mounting.

18-13 What are three important factors to consider when providing a seal?

18-14 Discuss a labyrinth seal.

18-15 What is the difference between a seal and a gasket?

18-16 Give several reasons why the actual motion produced by a cam may differ from the theoretical motion indicated by the cam profile.

18-17 Why should a designer become familiar with standard units?

18-18 What is the purpose of a Geneva drive?

PROBLEMS

Note: Upon completion of some of these problems, Design Example 7 and Project 16 in Chap. 19 are appropriate.

18-1 What starting torque is required to exert a 4500-lb force with a high-quality 2.5-in.-diameter Acme thread, if the outside diameter of the bronze thrust bearing is 3.5 and the inside diameter is 2.6?

18-2 What bending stress is induced in the outer wires of a 2-in. 6 × 19 rope in passing around a 5-ft-diameter sheave?*

18-3 What is the equivalent load due to bending a 1.5-in. 6 × 19 rope around a 4-ft-diameter sheave?

18-4 What force is developed in the upper end of a $2\frac{3}{4}$-in. 6×19 rope, 1200 ft long, hanging vertically, by accelerating only the rope at 20 ft/sec^2?*

18-5 What stress is induced in the upper end of a 6×19 rope, $1\frac{3}{4}$ in. in diameter and 1800 ft long, hanging vertically, by causing a point on it to move up 80 ft in 6 sec?

18-6 How long must a 1.5-in. 6×19 rope be to induce in the upper end a stress equal to one-fourth its ultimate strength if the rope hangs vertically?*

18-7 A flywheel rim weighs 200 lb, the initial speed is 400 rpm, the final speed is 200 rpm, and the mean diameter of the rim is 3 ft. How much energy is given up by the flywheel?*

18-8 What is the weight of the rim of a flywheel to be used with a dc generator which is to provide 800 ft-lb, if the average velocity of the center of the rim is 1500 fpm?

18-9 Determine the size for a cast-iron flywheel to be used with a reciprocating pump. The diameter is about 3 ft, and it turns 340 rpm. The energy to be supplied is 750 ft-lb. The rim is about 4 in. wide.

19 DESIGN EXAMPLES AND PROJECTS

INTRODUCTION

The importance of judgment and experience has been mentioned several times, and it has been pointed out that the beginning designer has little experience appropriate to design. The only way in which he can gain design experience is to design. The problems in the chapters have undoubtedly been worked, and it is essential for a designer to have the ability to solve such problems, but this is not design.

It is the purpose of this chapter to provide opportunities to exercise judgment and to gain experience that is as close to actual design experience as possible. It would be well to explain why the phrase "experience that is as close to actual design experience as possible" rather than "actual design experience" was used. In order that the desired experience be gained, the statement of the situation must be rather limiting. In other words, to avoid a long detailed description of the situation that would lead to the decision to use a particular machine element, the element to be used is specified. In actual design the many principles, concepts, and considerations, the discussion of which constituted the first half of this text, determine the particular elements and the arrangement to be used, also many of the details that must be incorporated.

The design examples require the exercise of judgment to a greater degree than do the examples in the various chapters, and the same is true of the projects when compared with the problems. The examples, problems, design examples, and projects are arranged to progressively develop ability in mechanical design.

The problems and projects that appear easy should be done conscientiously, for they are the basis for the more complex projects. In doing them it is likely that something will be learned or will become clear that the designer must understand before he can deal successfully with the more complex project.

The work done on a project should be presented as a report that includes the following:

1. All calculations arranged in a manner that can be easily followed
2. A statement supporting each assumption and decision that is made
3. Sketches that clearly illustrate the design and, where necessary, identify dimensions
4. A layout when appropriate
5. Recommendations so stated that it is clear on what basis they are made

PROJECT 1

Analyze the assembly shown in Fig. 19-1 to determine whether a tolerance stack exists. The details of the parts for this assembly are shown in Fig. 19-2. If a tolerance stack exists, assume that the tolerances on the parts cannot be changed and make a recommendation.

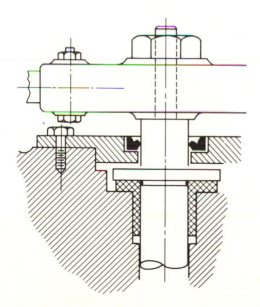

Fig. 19-1 Project 1.

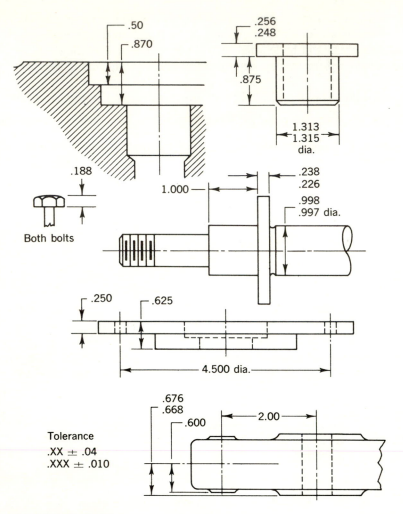

Fig. 19-2   Project 1.

PROJECT 2

A sand casting of aluminum alloy 356 is to be assembled to a shaft and, to avoid eccentricity caused by clearance, a shrink fit is desired. The shaft has a nominal diameter of 3 in. Design the joint. Include the recommended temperature and make a check of the stress induced.

PROJECT 3

Design a knuckle joint for a static tension load. The rod is 1015 hot-finished steel and is to have one of the following diameters: 1⅛, 1½, 1¾,

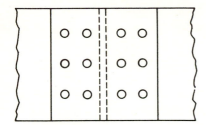

Fig. 19-3   Design Example 1.

2, 2¼, or 2½. The minimum strength of the joint is to be 10 percent greater than the strength of the rod in tension.

DESIGN EXAMPLE 1

Design a riveted butt joint in accordance with the AISC Code to transmit a 100,000-lb static load. The width of the plates is 15 in., the thickness is 0.5 in., and the holes are to be drilled.

Solution   The approximate rivet size is the square root of 0.5 or 0.707. Thus, the rivets selected for trial are: ¾, ¹³⁄₁₆, and ⅞. The shear strength and bearing strength are compared, as shown in Table 19-1, keeping

TABLE 19-1   Comparison of Shear Strength and Bearing Strength

RIVET DIAMETER	SHEAR STRENGTH	BEARING STRENGTH	DIFFERENCE
¾	13,300	15,000	1700
¹³⁄₁₆	15,500	16,200	700
⅞	18,000	17,500	500
¹⁵⁄₁₆	20,700	18,800	1900

in mind that the shear strength must be multiplied by 2, because the rivet is in double shear. The ¹⁵⁄₁₆ rivet was included because the ⅞ had the smallest difference and a larger rivet may have had a smaller difference. The ⅞ rivet appears to be the best. The number of rivets required is 100,000/17,500 = 5.7, or 6.

If the joint is arranged as shown in Fig. 19-3, the net width is 15 − (3 × 0.938) = 12.19 in., the net area is 12.19 × 0.5 = 6.10 sq in., and the tension stress in the net area is 100,000/6.10 = 16,400 psi. This stress is less than the allowed 20,000 psi; therefore, the arrangement is satisfactory.

The thickness of the cover plates is 0.266, the edge distance is 1.75, and the distance between the rows of rivets is 1.87.

PROJECT 4

Design a riveted lap joint to transmit a static load of 75,000 lb in accordance with the AISC Code. The plates are 0.375 thick and the holes are punched. The width is to be the minimum possible with two rows of rivets.

PROJECT 5

Design a welded joint for the situation shown in Fig. 19-4a. The joint is part of the truss of a crane. The location of the centroidal axis of the angle is shown in Fig. 19-4b.

DESIGN EXAMPLE 2

Design a shaft to transmit 12 hp at 600 rpm. The length is 9 in. There is a concentrated load of 90 lb at the center, suddenly applied. The torque is also suddenly applied.

Solution Weight is not a factor, and low-carbon steel should therefore be used. There will be little machining, and slight distortion will not be objectionable. Therefore, 1015 cold-drawn steel will be used. The

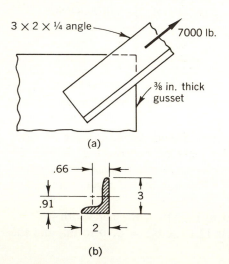

(a)

(b)   **Fig. 19-4   Project 5.**

properties from Table 3-3 are: tensile strength 56,000 and yield strength 47,000. E is 29,000,000 and G is 11,000,000.

The shaft is subjected to combined loading; therefore, strength factor b will be omitted from Eq. (7-1) in finding the design factor. Appropriate values for the other strength factors are: $a = 1$, $c = 1$ (fatigue and stress concentrations will be provided for elsewhere), $d = 1.2$. The design factor, therefore, is $1 \times 1 \times 1.2 = 1.2$. The appropriate strength of material to use in determining the design stress will be the endurance limit. The endurance limit per Sec. 7-4 is about 45 percent of the tensile strength. The value to use for torsion is one-half the endurance limit; thus, the appropriate strength of material is 12,600. The design stress found with Eq. (7-2) is $12,600/1.2 = 10,500$. An allowance must still be made for any stress concentrations. Since a keyway is to be used in the center of the shaft, the 10,500 should be reduced by 15 percent. The design stress thus becomes 8925.

The torque is found by rearranging Eq. (11-1):

$$T = \frac{63,000 \text{ hp}}{n} = \frac{63,000 \times 12}{600} = 1260 \text{ lb-in.}$$

Using the equation for a simple beam with a concentrated load in the middle, the maximum bending moment is

$$M = \frac{PL}{4} = \frac{90 \times 9}{4} = 202 \text{ lb-in.}$$

To allow for the load being suddenly applied (per Sec. 11-8) the factor $C_M = 1.6$, which increases the bending moment to 323. A factor must also be applied to the torque to allow for its being suddenly applied. From Sec. 11-8, $C_T = 1.2$. Thus, the torque is increased from 1260 to 1510.

The diameter of the shaft is determined by rearranging Eq. (11-9) and using the values just found for T, M, and the strength of material.

$$D^3 = \frac{5.1}{s_d} \sqrt{T^2 + M^2}$$

$$= \frac{5.1}{8925} \sqrt{1510^2 + 323^2}$$

$$D = 0.96 \text{ in.}$$

To check the deflection due to bending, the equation for the maximum deflection of a simple beam is used:

$$d = \frac{PL^3}{48\,EI}$$

$$= \frac{90 \times 9^3}{48 \times 29,000,000 \times 0.049 \times 0.96^4} = 0.00115 \text{ in.}$$

This is less than that which is acceptable.

To check the torsional deflection, Eq. (11-5) is used:

$$\theta = \frac{584LT}{GD^4} = \frac{584 \times 9 \times 1260}{11,000,000 \times 0.96^4} = 0.71°$$

The allowable for a machine shaft is 0.1° per foot. For a shaft 9 in. long, this would be 0.075°. Thus, the deflection is excessive, and the diameter of the shaft must be determined by the permissible torsional deflection. Rearranging Eq. (11-5),

$$D^4 = \frac{574LT}{\theta G} = \frac{584 \times 9 \times 1260}{0.075 \times 11,000,000}$$

$$D = 1.68$$

However, the diameter of the shaft must be changed to match standard bearings. Thus, the nominal diameter becomes 1.772 in. The diameter should be further increased at the hub of the mating part to reduce stress concentrations and to facilitate assembly if a press fit is used. If 1⅞-diameter stock was used, the nominal diameter at the hub could be 1.860 in., which would provide an adequate increase in diameter. To accommodate other details, such as shoulders for ball bearings, or threads, a slightly larger shaft may be required.

PROJECT 6

Design a rigid coupling for a 1¼-in.-diameter shaft, to transmit 16 hp at 1200 rpm. The torque load will be applied gradually, weight is not important, space is somewhat limited, and low cost is an important factor. The coupling will be used in an unenclosed space, and workmen will come close to it when it is operating. About 25 are to be made.

PROJECT 7

Design an overload-release coupling of the shear-pin type for the conditions given in Project 6, except that the coupling will be used in an enclosed space and the speed is 400 rpm.

DESIGN EXAMPLE 3

Determine the size of a main bearing for a centrifugal water pump to fulfill the following requirements: design load 4800 lb, 2800 rpm, maximum operating temperature 150°F, SAE 10 oil, nominal bearing diameter 5.00 in.

Solution The operating viscosity per Fig. 12-2 will be 12 centipoises. The velocity is found with Eq. (12-2):

$$V = 0.262DN = 0.262 \times 5 \times 2800 = 3670 \text{ fpm}$$

This velocity limits the choice of material to babbitt or silver. And as corrosion should not be a factor in a water pump, lead-base babbitt would be an appropriate material. Since the operating temperature is less than the maximum allowed, this is satisfactory.

To keep the bearing as small as possible, the minimum ZN/p should be used, which, according to Table 12-3, is 200. With ZN/p equal to 200, Z equal to 12, and N equal to 2800, the bearing pressure p is found to be 168 psi. This is less than the maximum pressure specified in Table 12-3 and much less than the maximum for lead-base babbitt given in Table 12-2. The projected area of the bearing is, therefore, $4800/168 = 28.6$ sq in. The bearing length is then $28.6/5 = 5.7$ in. The L/D ratio is 1.14, which is slightly larger than the minimum specified in Table 12-3 and is, therefore, satisfactory. In completing the design, attention must be given to the details covered in Sec. 12-16.

PROJECT 8

Design a wrist-pin bearing for a reciprocating compressor. The design load is 5000 lb, the rpm is 3800, and the maximum operating temperature is 275°F. SAE 20 oil is to be used, and the nominal diameter is 2.00 in. Length is limited.

PROJECT 9

Design a bearing to take a radial load of 360 lb and a thrust load of 25 lb. The journal turns 800 rpm. The bearing is located where frequent attention is undesirable, replacement would be very inconvenient, the total operating life will be only 200 hours over a period of several years, the environmental conditions are not severe, and no special conditions exist. The journal is to have a nominal diameter of 2⅞ in.

DESIGN EXAMPLE 4

Design a V-belt drive to transmit 16 hp under average conditions. The driving pulley turns 2000 rpm, and the driven pulley is to turn 880 rpm. The distance between centers is somewhat limited, the belt should have a reasonable life, and efficiency is important.

Solution Referring to Fig. 14-7, and considering the space, life, and efficiency requirements, it appears that a *B*-belt would be best for the first trial. A 7-in.-diameter pulley is used as the driver. The velocity is found with Eq. (14-4):

$$V = 0.262DN = 0.262 \times 7 \times 2000 = 3680 \text{ fpm}$$

The diameter of the large pulley is found with Eq. (14-2):

$$2000 \times 7 = 880 \times D_2$$

$$D_2 = 16 \text{ in.}$$

To keep the size to a minimum, a center distance of 18 in. is chosen. This is slightly larger than the diameter of the larger pulley. The angle of contact is found with Eq. (14-8):

$$\theta = 57.3\left(\pi - \frac{D-d}{c}\right) = 57.3\left(\pi - \frac{16-7}{18}\right) = 152°$$

This is greater than 120° and is, therefore, satisfactory. From Fig. 14-7 it is determined that a B-belt will transmit about 6.8 hp at a velocity of 3680 fpm. For average duty, the factor is about 1.5. Thus, the 6.8 is divided by 1.5, which is 4.5. This is the corrected capacity of the belt and is used to determine the number of belts required, which is $16/4.5 = 3.56$, or 4 belts.

PROJECT 10

A drive is to transmit 50 hp. The driver turns 1750 rpm, and the driven pulley is to turn 750 rpm. The distance between centers is limited, the components should have a reasonable life, efficiency is important, and light shock is involved. Make calculations for three of the following: flat belt, V-belt, roller chain, and silent chain. Determine the space required, and comment on the advantages and disadvantages of each type. Also make a recommendation of the type to use.

DESIGN EXAMPLE 5

Design a gear drive for the following situation: 50 hp is to be transmitted, the pinion turns 1750 rpm, the required speed of the gear is about 550 rpm, the diameter of the pinion shaft is 2 in., quiet operation is not required, cost and size are of moderate importance, intermittent operation is required for a total of several hours per day, and there is little uncertainty and no shock.

Solution Since size is of moderate importance, the gear will be keyed to the shaft rather than integral with it, and a 5-in. diameter will be used (this is more than twice the diameter of the shaft — see Sec. 15-6).

The velocity is found with Eq. (15-3):

$$V = 0.262DN = 0.262 \times 5 \times 1750 = 2300 \text{ fpm}$$

The force on the tooth is found with Eq. (15-2):

$$F = \frac{33,000 \text{ hp}}{V} = \frac{33,000 \times 50}{2300} = 720 \text{ lb}$$

Because there is little uncertainty and no shock, this 720 lb will be compared with the allowable load.

A 5-pitch tooth size looks about right, and it would be desirable to use high-grade cast iron for the material. A 5-pitch 5-in.-diameter gear has 25 teeth, and the circular pitch is $\pi/P = 0.628$. According to Sec. 15-6, the face width should be 3 to 4 times the circular pitch. If 3.5 is used, the width is $3.5 \times 0.628 = 2.2$ in. The tooth form factor from Fig. 15-6 is 0.34. The allowable stress from Table 15-1 is 13,000 psi. The allowable load is found with Eq. (15-4):

$$F_s = \frac{sfY}{P} = \frac{13,000 \times 2.2 \times 0.34}{5} = 1940 \text{ lb}$$

The factor K must be applied to this load before it can be compared with the 720 lb. The factor K for accurately hobbed gears at velocities less than 4000 fpm (which is appropriate for this situation) is found with Eq. (15-6):

$$K = \frac{1200}{1200 + V} = \frac{1200}{1200 + 2300} = 0.343$$

The corrected allowable load is thus $0.343 \times 1940 = 665$ lb. This is less than 720 lb and is, therefore, not satisfactory. The face width could be increased to 4 times the circular pitch:

$$4 \times 0.628 = 2.51$$

The allowable load is

$$F_s = \frac{13,000 \times 2.51 \times 0.34}{5} = 2210$$

Applying the factor K,

$$0.343 \times 2210 = 758 \text{ lb}$$

This is greater than 720 lb and is, therefore, acceptable.

If increasing the width had not provided sufficient strength, making the pinion of steel and the gear of cast iron could be tried. It would then be necessary to check the strength of the gear — the pinion would obviously be strong enough. Using 6-pitch gears would also be an appropriate trial.

The required ratio is $1750/550 = 3.19$. Thus, the number of teeth on the gear should be $3.19 \times 25 = 79.5$, which must be 80 teeth. The speed of the gear will be 547 rpm. Since the 550 rpm was approximate, this should be close enough. The diameter of the gear is found by dividing 80 by 5 (the pitch); thus, the diameter is 16 in. The center distance is one-half the sum of the diameters of the two gears, that is, $2.5 + 8 = 10.5$ in.

PROJECT 11

Design a gear drive for the following situation: 30 hp is to be transmitted, the pinion turns 1160 rpm, the required speed of the gear is 375 rpm, the diameter of the pinion shaft is 1.5 in. Quiet operation is not required, a small size is more important than a low cost, operation is intermittent for a total of several hours per day, and there is a small amount of uncertainty and slight shock.

PROJECT 12

Design a gear drive for the situation specified in Project 11 except that quiet operation is required and size is not important, but low cost is.

PROJECT 13

A cast-iron cone clutch is required for use in a gear case where the gears are lubricated by an oil spray. The operation is only for short periods and is accomplished by the operator's pulling on a lever. He should not have to move the lever more than 8 in., nor exert more than a 10-lb force. The 3 hp that the clutch is to transmit when turning 800 rpm includes an allowance for all appropriate strength factors. Assume that the housing can be arranged to provide a pivot point for the lever wherever desired. Ignore the friction between the lever and the clutch. Design only the clutch, but make provision for the actuating lever. Also assume that the lever will be arranged to keep the clutch disengaged except when operated.

DESIGN EXAMPLE 6

Design a compression spring for average service, to be made from hard-drawn wire, that will deflect 1.5 in. when subjected to a 300-lb load.

Solution   The factor for average service is 1.8, and the yield strength in shear of hard-drawn wire is 95,000; therefore, the allowable stress is $95,000/1.8 = 53,000$ psi.

Rearranging Eq. (17-4),

$$d = \sqrt[3]{\frac{8PDK}{s_s'\pi}}$$

Assume $D = 2$ in. and $K = 1.3$. Thus,

$$d = \sqrt[3]{\frac{8 \times 300 \times 2 \times 1.3}{53,000\pi}} = 0.335 \text{ in.}$$

The standard wire size 0.331 is chosen from Table 17-1. Check K: $C = D/d = 2/0.331 = 6.05$. For $C = 6.05$ in Fig. 17-3, $K = 1.25$, and therefore the assumed K of 1.3 is close enough.

Rearranging Eq. (17-2),

$$N = \frac{FGd^4}{8PD^3}$$

G for hard-drawn wire is 11,400,000. Thus,

$$N = \frac{1.5 \times 11,400,000 \times 0.331^4}{8 \times 300 \times 2^3} = 10.7 \text{ coils}$$

The minimum clearance at minimum working length is 10 percent of the wire diameter or 0.033. Thus, the total allowance for clearance is $0.033 \times 10.7 = 0.354$ in. The solid length for a spring with squared and ground ends is $d(N + 2) = 0.331(10.7 + 2) = 4.20$ in. Thus, the minimum working length is $4.20 + 0.354 = 4.554$ in., and the free length is $4.554 + 1.5 = 6.054$ in. Allowing for tolerances, the smallest practical free length should not be less than 6.25 in.

As K was used in determining the wire size, it will not be necessary to make any other allowance. It will not be necessary to consider surge for a spring intended for average service. As the length is considerably less than 5 times the mean diameter, buckling will not occur.

A check of the stress at the solid length should be made. The free length is 6.25, and the solid length is 4.20; the difference is 2.05. Rearranging Eq. (17-5) to solve for s_s and substituting 2.05 for F,

$$s_s = \frac{2.05 \times 11,400,000 \times 0.331}{2^2 \times 10.7\pi} = 57,700 \text{ psi}$$

This stress must be corrected by multiplying by K. Thus, the corrected stress is 72,000 psi, which is less than 95,000 and, therefore, satisfactory.

PROJECT 14

A mechanism involving a compression spring is to operate 3 to 5 times a day. The anticipated life of the mechanism is 15 years, and failure of the spring would not cause damage but would require one man to spend about one hour to replace it. There are no special problems because of the environment in which the mechanism is to operate; cost is rather important; and space is somewhat limited. The spring must deflect 2 in. when subjected to a 175-lb load.

Design the spring to find the outside diameter and the initial working length, for they affect the design of the mechanism. Then prepare a description of the spring to be submitted to the spring manufacturer.

PROJECT 15

A slip clutch is to be incorporated in a unit to prevent overload and will slip for very short periods several times a day. It is to be located in an enclosed space. It will operate dry, and no special environmental problems are expected. Weight is not important. Space along the shafts is somewhat limited, but radial space is not. Cost is rather important, and 50,000 units are to be produced. The clutch is to limit the power transmitted to 10 hp at 1800 rpm. The torque at which slipping is to occur is critical; therefore, an adjustment must be provided. One shaft has a diameter of 1 in., the other 1¼ in. Design the clutch and the ends of both shafts.

DESIGN EXAMPLE 7

Determine the size of a 6×19 wire rope for the following situation: a mine hoist is to lift 1 ton of ore 800 ft, the acceleration is to take place in 100 ft and 5 sec, the ore skip weighs 1500 lb.

Solution The solution is made by trial. Assume that three ½-in. ropes are to be used. The weight of the ropes is $3 \times 800 \times 1.6 \times 0.5^2 = 960$ lb. The total weight to be accelerated $= 2000 + 1500 + 960 = 4460$ lb. The force in the rope due to acceleration is found with Eq. (18-5):

$$F_a = \frac{2WL}{32.2t^2} = \frac{2 \times 4460 \times 100}{32.2 \times 5^2} = 1110 \text{ lb}$$

The equivalent bending load is found with Eq. (18-4):

$$F_b = \frac{AE_r d_w}{D}$$

$$= \frac{0.38 \times 0.5^2 \times 12,000,000 \times 0.063 \times 0.5}{65 \times 0.5} = 11,100 \text{ lb}$$

The total load, which is the working load in the rope, is as determined in Table 19-2.

The ultimate strength of the rope is $72,000 \times 0.5^2 = 18,000$ lb. The total strength of 3 ropes is $3 \times 18,000 = 54,000$ lb.

The design factor is, therefore, $54,000/16,670 = 3.24$. The recommended design factor for a mine hoist is 3; thus, three ½-in. ropes is satisfactory.

TABLE 19-2 Total Load

Weight of ore	2000
Weight of skip	1500
Weight of ropes	960
Allowance for acceleration	1110
Equivalent bending load	11,100
Working load	16,670

PROJECT 16

A 15-passenger elevator is to have a lift of 1500 ft, the acceleration is 75 ft in 5 sec, the cage weighs 2500 lb, 6 × 19 rope is to be used, the design factor is 10, and 2 tail ropes are to be used. Determine the size of the hoist ropes and the tail ropes.

GLOSSARY

Terms discussed in the text are listed in the Index.

alloy   A mixture of two or more elements, at least one of which is a metal.

anneal   To heat and cool a metal gradually in order to increase its ductility.

anodize   To form a corrosion-resistant coating by a process similar to plating. Used most frequently with aluminum.

arbor   A shaft or axle.

axial load   A load acting along the axis of a part.

boss   A cylindrical projection on a casting or forging, usually intended as a seat for a bolt head or nut.

brake forming   Forming metal in a brake. A brake is a machine used for bending sheet through various angles.

brazing   Joining of two pieces of metal by use of a molten filler metal. The pieces being joined are not melted.

Brinell hardness number (Bhn)   A method of testing the hardness of a metal. A steel ball is forced into the metal and the resulting impression is measured.

broaching   Removing metal with a broach. A broach is a cutting tool with a series of teeth that gradually increase in size. It is pulled through a hole or over a surface to produce the cutting action.

cantilever beam   A beam that is supported at only one end.

carbon steel   Steel to which no special alloying elements have been added.

case hardening   The hardening of the outer surface of a ferrous alloy by heating and quenching. The carbon content of the surface must be increased in some manner.

center of gravity   The theoretical point at which the weight of a body appears to be concentrated. The center of gravity of an area would be the same point as the center of gravity of a sheet of metal of constant thickness which has the same configuration as the area.

chuck   A mechanism for holding a rotating workpiece or cutting tool.

coin   To press a cold metal part between dies to produce the desired size. The surface roughness can also be improved by coining.

cold rolling   Cold working a metal by rolling.

cold working   Deforming a metal, at about room temperature, to produce the desired shape.

column   A slender member subjected to a compression load.

compression   Forces acting to shorten a body produce compression.

core   A form, usually made of sand, which is placed in a mold to produce a configuration that cannot be achieved by use of just a pattern.

core shift The amount that the surfaces of a casting produced by the core are displaced relative to the surfaces produced by the pattern.

creep The increase in strain, with time, of a body under stress. The amount of creep increases with an increase in temperature.

cupola A vertical furnace for melting metal in which the fuel and the metal are in contact.

die casting A process of forcing molten metal into metal dies to produce an accurate, smooth casting.

embrittlement The reduction in the ductility of a metal because of a physical or chemical change.

fillet The inside radius between two intersecting surfaces.

flame hardening The hardening of a metal in which the heat is applied directly by a flame.

frequency The number of cycles of a frequent occurrence completed in a specified time.

gage length The original length of a portion of a specimen used in the determination of strain and elongation.

grain (1) The crystals in a metal. (2) The arrangement of the structure of a metal so as to have a directional character (similar to wood). The result of rolling or forging.

heat-treating A process of heating and then cooling a metal to obtain the desired properties.

helical Having the form of a spiral.

hob A rotary cutting tool with teeth arranged along a helix that is used to produce a toothlike form. Both the workpiece and the hob must be rotated while the hob is fed across or into the workpiece.

honing The removing of a small amount of metal by use of abrasive sticks mounted in a holder which is moved relative to the workpiece.

hot rolling Deforming a metal, by rolling, to the desired shape while it is hot enough to be in the plastic state.

hydrogen embrittlement Embrittlement resulting from the absorption of hydrogen.

impermeable Not permitting things to pass through.

intergranular corrosion Corrosion occurring at the boundaries of the crystals that make up the metal.

investment casting The process in which a mold is produced by coating the pattern with the mold material, then removing the pattern by melting it.

kinematic Pertaining to the geometry of motion.

kinetic energy The energy that a body has because of its motion.

lapping A finishing operation that makes use of a lap. A lap is an appropriately shaped object upon which fine abrasive is placed.

linear velocity Velocity measured along a straight line.

mismatch The error in register between surfaces formed in opposing dies or molds.

modulus of elasticity The ratio of tensile stress, below the proportional limit, to the accompanying strain.

modulus of rigidity The ratio of torsional stress, below the proportional limit, to the accompanying strain.

mold A form which has a cavity into which molten metal is poured to make a casting.

oxidation To combine with oxygen. Rust is such a process.

pad A slight projection on a casting or forging which is generally machined to provide a smooth, accurate seat.

parting plane The plane that separates the two parts of a mold.

percentage of elongation In tensile testing, the increase in the gage length, measured after fracture, expressed as a percentage of the original gage length.

permanent-mold casting A casting process employing metal molds. The molten metal flows by gravity, rather than being forced into the mold.

proportional limit The maximum stress at which the strain is proportional to the stress.

prototype An early experimental model of a product that is as close to the proposed production model in every detail as possible.

quench To cool a piece of metal rapidly. Part of the heat-treating process.

radian A measure of an angle. One radian equals approximately 57° 18′.

residual stresses Stresses produced by a manufacturing process that remain after the part is completed.

resistance welding A welding process that involves electrical heating and pressure.

Rockwell C A method of testing the hardness of a metal. A penetrator is forced into the metal and the depth of penetration is measured.

scoring Marring of a surface.

seizing The stopping of motion between two parts as the result of galling.

serrations Shallow toothlike projections produced on a surface.

shim A thin piece of material used as a spacer to adjust two parts.

static friction Friction that must be overcome to start motion. It is greater than running friction, which is the friction to be overcome to continue motion.

stamping The process of producing parts from sheet metal by the use of dies placed in a press.

strain Strain in tension testing is the total elongation in inches divided by the original gage length in inches.

strain-harden To harden by cold working.

stress The force acting on a body divided by the cross-sectional area that resists it.

stress relief Reducing residual stresses by heating and cooling.

superfinishing A process similar to honing. The abrasive stones are spring-loaded.

tapping Cutting internal threads with a tool called a tap.

temper (1) To change the physical characteristics of a metal by heat-treating. (2) The hardness and strength of a metal produced by cold working.

tension Forces acting to lengthen a body produce tension.

thread rolling Threads produced by deforming the metal rather than by removing part of it.

transverse shear The tendency of a beam to shear in a plane parallel to the load.

ultimate strength The ultimate strength of a material is the maximum stress which a specimen can withstand. Unless otherwise stated, it is the tensile strength.

wrought Wrought metal has been worked in some manner, such as by rolling.

yield strength The stress at which the stress-strain curve deviates from a straight line by a specified amount called the offset. An offset of 0.2 percent is often used.

ANSWERS TO SELECTED PROBLEMS

A problem for which an answer is given is indicated by an asterisk in the text.

CHAPTER 8

8-1 0.004 to 0.010 clearance
8-3 0.0007 clearance to 0.0002 interference
8-5 0.6243/0.6235
8-7 75,000 psi
8-9 430°F

CHAPTER 9

9-1 3000 lb-in.
9-3 67° (about ⅕ turn)
9-7 0.857
9-9 0.176

CHAPTER 10

10-1 10,150 psi shear; 9150 psi bearing
10-4 1.0 in.
10-9 0.83

CHAPTER 11

11-1 114 hp
11-2 4030 psi
11-5 319 psi shear, 510 psi tension
11-9 13.75 lb-in.

CHAPTER 12

12-1 180
12-3 65.7
12-5 197
12-8 23,400

CHAPTER 13

13-1 6.6 million revolutions
13-3 No. 12 bore
13-5 4050 lb
13-8 13,300 lb
13-9 2860 lb

CHAPTER 14

14-1 320 rpm
14-3 1675 fpm
14-4 149°
14-7 163.7
14-9 146 links (73 in.)

CHAPTER 15

15-1 6, 4.33, 3.614, 0.524, 0.360
15-2 467 lb
15-5 12
15-8 43.6 lb

CHAPTER 16

16-1 150 lb
16-3 4.58 hp
16-5 1440 lb-in.

CHAPTER 17

17-1 165,000 psi
17-3 0.0626
17-5 91,700 psi
17-9 160,000 psi, 20 in.

CHAPTER 18

18-2 25,200 psi
18-4 9000 lb
18-6 9640 ft
18-7 97.6 ft-lb

INDEX

INDEX

This book was set in Times Roman by Trade Composition, Inc. and printed on permanent paper and bound by The Maple Press Company. The designer was Edward Zytko; the drawings were done by Bertrick Associate Artists, Inc. The editors were Cary F. Baker, Jr. and Albert Shapiro. Morton I. Rosenberg supervised the production.